T0074623

A Pocket Guide to Scientific Writing in Aquaculture Research

A Pocket Guide to Scientific Writing in Aquaculture Research

Rodrigue Yossa
WorldFish, Penang, Malaysia &
AquExperts International Inc.,
Ottawa, Ontario, Canada

CRC Press
Taylor & Francis Group
Boca Raton London New York

CRC Press is an imprint of the
Taylor & Francis Group, an **informa** business

First edition published 2022
by CRC Press
6000 Broken Sound Parkway NW, Suite 300, Boca Raton, FL 33487-2742

and by CRC Press
2 Park Square, Milton Park, Abingdon, Oxon, OX14 4RN

CRC Press is an imprint of Taylor & Francis Group, LLC

Library of Congress Cataloging-in-Publication Data
Names: Yossa, Rodrigue, author.
Title: A pocket guide to scientific writing in aquaculture research / Rodrigue Yossa.
Description: First edition. | Boca Raton, FL: CRC Press, 2022. |
Includes bibliographical references and index. |
Summary: "Writing a high-quality scientific aquaculture publication is challenging, and many students and early career aquaculture scientists find the task daunting. Yossa provides new researchers with all the tools they need to write abstracts and a variety of articles (peer-reviewed, technical, magazines, working papers, scientific reports and more). He talks the reader step-by-step through the process of reviewing submitted manuscripts and replying to reviewers, and understanding research ethics. Each section is accompanied by examples, offerings a lifeline to aquaculture students and early career academics getting to grips with the basics"— Provided by publisher.
Identifiers: LCCN 2021036368 (print) | LCCN 2021036369 (ebook) |
ISBN 9780367338879 (paperback) | ISBN 9780367338886 (hardback) |
ISBN 9780429322648 (ebook)
Subjects: LCSH: Aquaculture—Research—Handbooks, manuals, etc. |
Technical writing—Handbooks, manuals, etc. | Handbooks and manuals.
Classification: LCC PE1475 .Y67 2022 (print) | LCC PE1475 (ebook) |
DDC 808.06/6—dc23
LC record available at https://lccn.loc.gov/2021036368
LC ebook record available at https://lccn.loc.gov/2021036369

ISBN: 978-0-367-33888-6 (hbk)
ISBN: 978-0-367-33887-9 (pbk)
ISBN: 978-0-429-32264-8 (ebk)

DOI: 10.1201/9780429322648

Typeset in Palatino
by codeMantra

This book is dedicated to Gaelle Yossa

Contents

Foreword

Publish or Perish!

One of my mentors told me many years ago that if I wanted to succeed as an aquaculture scientist, I had to "publish, publish, publish". Publishing is the bread and butter of the scientific community. Most of us do research to solve a problem, but problems won't get solved if no one knows about the research. Peer-reviewed publications make our research available to those who need it. Publications also pave the way to a successful career in science.

Sadly, too many scientists, especially scientists working in underfunded laboratories in developing countries, struggle to get their research into the public domain. Sometimes it's a quality issue resulting from a lack of the ever more expensive equipment and staffing needed to engage in the cutting-edge science that is attractive to the frontline journals. Many times, however, it is just the way a manuscript is prepared and presented that results in rejection. This book is aimed at correcting that problem.

Dr. Rodrigue Yossa, a native of Cameroon in Central Africa, has been through all of this and comes now to advise younger aquaculture scientists on how to avoid the pitfalls of peer review. It's not easy. Reviewers sometimes have a bad day and take it out anonymously on good junior scientists because of a messy and disorganized manuscript. It can be very disheartening to work hard on a research project and have it ignored because of the way it was presented.

Read this book, take these lessons and apply them to your next manuscript and I'm sure you'll find a warmer reception among the international aquaculture community.

Randall E. Brummett

Preface

As the Editor-in-Chief of the *Journal of Applied Aquaculture*, I have experienced firsthand how difficult it is for many students and inexperienced aquaculture scientists to write aquaculture manuscripts. While I am of the opinion that there is no acceptable justification for poor writing skills, I have come to realize that not every aspiring aquaculture scientist (writer) has access to the training and exposure that are required to gain and strengthen their scientific writing skills. Considering my own academic path, which involved studying in Cameroon, Belgium, The Netherlands, Vietnam, Canada and the USA, I can say that being taught by professors who actively publish scientific articles can substantially contribute to inspiring and motivating aspiring scientific writers. In addition, studying at universities where the library (physical and online) puts at the disposal of students and researchers tones of literature allows them to quickly acquire writing skills and tools.

Given that not everyone will have the chance to follow the same path as I and learn how to write scientific manuscripts through international studies, editorial work and professional experience across the world, I thought I would disseminate my insights with anyone who is wondering how to write aquaculture publications. My journey on the sharing of scientific writing skills started in 2014, when I published the paper titled "Writing a scientific manuscript from original aquaculture research" in the *Journal of Applied Aquaculture*. I then followed up in 2015 with other papers titled "Toward the Professionalization of Aquaculture: Serving as Peer Reviewer for an Academic Aquaculture Journal" published in *World Aquaculture* magazine and "Misuse of multiple comparison tests and underuse of contrast procedures in aquaculture publications" published in *Aquaculture*. Furthermore, I was invited to deliver a training workshop on "Improving scientific writing and tools for research organization" at the World Aquaculture Society annual meeting and conference in Cape Town, South Africa, in 2017. All the experience gained through these science communication activities are collated in this book, in order to better equip the readers with the tools and methods required to write quality aquaculture abstracts and a variety of manuscripts (original, research reports, magazines, working papers, conference proceedings and more) and increase their chances of getting their aquaculture findings into print.

Rodrigue Yossa
Penang, Malaysia

Acknowledgments

I would like to thank Dr. Randall Brummett for giving me the first opportunity to work with living fish in Cameroon in 2002 and for offering me the opportunity to serve as Managing Editor of the *Journal of Applied Aquaculture* under his leadership between 2012 and 2015. I would also like to express my gratitude to my family, friends, co-authors, colleagues, and teachers and professors (Patrick Sorgeloos, Peter Bossier, Gilbert Van Stappen, Jean Dhont, Grant W. Vandenberg, Marc Ekker, Jean François Bernier, Jancice L. Bailey, Johan Verreth, Marc Verdegem, Céline Audet, Jonathan C. Eya, Victor Pouomogne, Steve Sulem, David Nguenga, Minette Tomedi Eyango, Jules Jacques Antoine Kome and Tchanou Zachée) who continually help me to become a better aquaculture specialist, scientist, communicator, farmer and, above all, a better person. I am grateful to all the referees who have reviewed my manuscripts, including the reviewers of the current book, for their constructive remarks and guidance. I thank my publisher, Taylor & Francis Group, for guiding me during the writing of this book and for publishing it. I finally thank you, the readers, for taking the time to go through this pocket guide.

Author

Rodrigue Yossa is an Aquaculture Specialist with 18 years of aquaculture research, development and program and project management experiences in North America, Asia and Africa. He is the Editor-in-Chief of the *Journal of Applied Aquaculture* and an aquaculture scientist at WorldFish, which is an international research for development organization member of the One CGIAR system. Prior to joining WorldFish, He was the Scientific Director of the Aquaculture Division at the Coastal Zones Research Institute Inc. (now Valores), in New Brunswick, Canada, where he led a team of 12 aquaculture scientists and technicians. He was also an Adjunct Professor at the University of Moncton in Canada, from 2016 to 2019.

Rodrigue Yossa is the co-founder of AquExperts International Inc. (www.aquexperts.com), which is a Canada-based international group that provides scientific, professional and technical services in aquaculture, fisheries and management of aquatic resources in Africa. He is a certified Project Management Professional (PMP) from the Project Management Institute. He holds a PhD in animal sciences from Université Laval (Canada), an MSc in aquaculture from Ghent University (Belgium), University Certificates in Biotechnology from Université Laval and in art and philosophy from Ghent University and a BSc in water, forestry and wildlife engineering from University of Dschang (Cameroon). He is fluent in English, French, Bafang language, and Pidgin English spoken in Cameroon, Nigeria,

Ghana, Sierra Leone and Liberia. He is also an advocate of sustainable organic farming. He has visited more than 30 countries across the world, for aquaculture research and development purposes.

1

Introduction

To write or not to write?

Although publishing is the bread and butter of any scientist, writing a publishable scientific manuscript is a struggle for many. The reality is that writing is essential in any scientific discipline, including aquaculture science, because "If it is not written down, it didn't happen". Moreover, publishing observations, facts and findings legitimizes a creation and establishes propriety. Every other person using the idea is bound to acknowledge the source of this idea by citing its author. Publication also defines productivity. Authors who publish many articles are regarded as "highly productive" and more likely to receive promotions and other accolades.

Writing combines inspiration, patience, imagination and hard work. Inspiration is subjective, as people get their inspiration from diverse sources reading, isolation, meditation, prayers, drinking, etc. Writing is time-consuming and involves many rounds of rewriting, editing and review, involving blood, sweat and tears. Thomas Edison said "Genius is 1% inspiration and 99% perspiration" and "Many of life's failures are people who did not realize how close they were to success when they gave up". Imagination is the result of deep thinking, vision and/or creativity. Albert Einstein once said that "imagination is more important than knowledge". Imagination is necessary during the identification and expression of the research question, during the design of the experiment, including the definition of treatments and

DOI: 10.1201/9780429322648-1

setting of the experiment (upstream), and during the writing of the manuscript, especially in the discussion section (downstream).

Writing an aquaculture manuscript is the last activity in the scientific research cycle, as it only happens once the experiment and the hard work that is related to the execution of the aquaculture experiment on-station or on-farm is completed, and the results are deemed relevant to the end users. It is worth noting that the end users of the message contained in an aquaculture article are not necessarily the readers of the articles. The readers could be farmers, academics, extension services, policy makers, government officers or any other person who can read, understand the message and then relay it directly or indirectly to the farmers, feed millers, etc. for use.

Although there are many types of publications, including peer-reviewed articles, research report, working papers, conference abstract, articles in conference proceedings and technical articles published in magazines, which will all be covered in this book, the peer-reviewed articles are still considered as the highest ranking publication resulting from a single experimentation. The peer review is the process used by editors and publishers of academic journals to invite independent scientists (peer reviewers) to evaluate manuscripts submitted for publication. A successful peer review of a manuscript leads to its acceptance and publication, while an unsuccessful peer review leads to its rejection and therefore no publication. Peer review is an important way to spread significant advances in knowledge in the aquaculture community because the quality of the overall scientific work, presentation of research and results and novelty of findings are validated by peers prior to publication. Peer review publication is also the process through which scientific research is given legitimacy. The number of published peer-reviewed articles, the quality of the journal where they have been published and the number of article citations are commonly used by universities and research institutions in the evaluation of scientific productivity of an aquaculture scientist or researcher (Hayer et al. 2013). Along with the rise of the aquaculture industry, the number of international scientific journals dealing with aquaculture has increased from less than ten just a few decades ago to over 130 nowadays.

Aquaculture is predominantly an affair of less-industrialized economies. Despite the dominance of, particularly, Asia in terms of fish production and consumption, the technology that drives productivity and efficiency is derived from scientific research that is largely conducted and published in Europe and North America. Less than 20% of the world's population lives in the 34 member countries of the Organisation for Economic Co-operation and Development (OECD), but out of the total of 139,257 engineering and technology papers published in 2007, almost 60% came from OECD countries (UNESCO 2010). However, over the last decade, we have seen more and more published papers whose authors are based in Asia. For instance, the science and engineering publication output of China rose by fivefold between 2003 and 2016, and China output in terms of absolute publication quantity is now comparable to that of the United States (NSF 2017).

1.1 Why Writing Matters

Authors write scientific papers for many reasons, including the willingness to share their knowledge, personal pride, recognition, financial incentive, institutional visibility, graduation, promotion, job applications and fund raising. Some authors are very happy to dedicate their time and energy to sharing the extensive experience that they have acquired on one or several subjects over time with a large audience through articles or books. The best example who comes to my mind while writing this book is Professor Claude Boyd of Auburn University, who is a regular contributor on topics related to water and pond soil not only in peer review journals, books and book chapters but also in popular aquaculture magazines. Some authors are proud to have their names attached to publications, to be recognized as published authors and to have their names appearing in online search engines with a link to a published title that they have authored or co-authored. Some professional aquaculture authors make a living through writing and receive a pay check for every article that they contribute to aquaculture journals or magazines; whereas, other authors receive financial incentives in the form of bonuses every time they publish peer-reviewed articles, book chapters or books, as is the case in some African, South American and Asian universities and research organizations where publications improve the image and visibility of the employer (organizations). A certain number of publications is required for graduation from academic institutions especially at the graduate level (MSc and PhD) and for professional promotion. The expression "publish or perish" describes well the pressure of publishing articles that is on the shoulders of the scientific community. Some universities or academic advisors ask students to publish at least one article for an MSc and two for a PhD, as a minimum requirement for graduation. The promotion from Associate Scientist to Junior Scientist to Scientist to Senior Scientist and to Principal Scientist, and from Assistant Professor to Associate Professor to Professor are all associated with a certain number of publications that demonstrate the scientific and academic maturity of the applicant. Furthermore, the number and the quality of publications represent an important criterion when selecting candidates for scholarships, university admission and aquaculture jobs. Many scientists are scored on the basis of their scientific publications during the evaluation of grant proposals by the funding agencies. The more prolific authors, who have authored or co-authored many referred articles, would thus score high and have more chances to have their grant proposals funded.

Despite the popularity of scientific publications, it is still considered inappropriate to ask a scientist or a professor how many articles she/he has published, because there is no established threshold on the number of publications that can be considered as good or bad. For the same reason, it is considered arrogant to boast about the total number of articles that you have published. It is however acceptable to create a blog or webpage or ResearchGate account or Google scholar account where you compile your publications, because only people who are interested in your scientific productivity will be inclined to check your publication list.

1.2 Who Should Write Publications

Everyone is authorized to submit peer-reviewed or non-peer-reviewed aquaculture manuscripts for publication, irrespective of one's race, academic qualification, professional affiliation, work experience, religion, country of origin, political aspirations, mother tongue, etc., as long as the content of the submitted manuscript is aligned with the scope of the receiving journal and the text is edited as per the instructions for authors, available on the journal's website. The focus of the review process is not the authors but rather the material (manuscript, cover letter, etc.) submitted to the receiving journal for review and publication.

It is thus important that the topic covered by the manuscript contains original information, advances knowledge on the topic and presents the results generated through a rigorous scientific approach.

Some journals have a scope that is specific to one research area, such as *Aquaculture Nutrition* and *Aquaculture Economics & Management*, while other journals are more general and accept papers covering a wide range of aquaculture topics, including *Aquaculture, Aquaculture Research, Journal of Applied Aquaculture, Journal of the World Aquaculture Society*, etc. Journals usually accept manuscripts written in only one language, with English being the most popular, although some journals still receive manuscripts written in French (*Tropicultura, Cahier de la Méditerranée* and *Aquatic Living Resources*), Spanish (*AquaTIC, Acta Agronómica, Limnetica* and *Boletín del Instituto Español de Oceanografía*), Portuguese (*Acta Fisheries and Aquaculture, Acta Scientiarum. Animal Sciences, Boletim do Laboratório de Hidrobiologia, Acta Iguazu* and *Arquivos de Pesquisa Animal*), German (*Archives Animal Breeding*), Japanese (*Aquaculture Science* and *Nippon Suisan Gakkaishi*), Korean (*Korean Journal of Fisheries and Aquatic Sciences* and *Journal of the Korean Society of Fisheries and Ocean Technology*), and Mandarin (*Journal of Fishery Sciences of China, Fisheries Science & Technology Information* and *Fisheries Science*). English is popular in the scientific aquaculture community, as it is generally considered to be the "language of science". For most of the scientists in the world, English is not their first language, but this should not prevent them from being widely published. Scientific writing tools and techniques are the same irrespective of the language because scientific principles are rigorous, not flexible and not sensitive to the scientist's mother tongue.

The instruction for authors dictates the way the main text, figures, tables and artwork must be edited and presented prior to the submission of the aquaculture manuscript. The instructions for authors provide information on the scope of the aquaculture journal, the type of English to use (e.g., American English versus British English), the main sections of the manuscript, the reference style to be used both in the text and the reference list and the format of the units of measure such as the metric system, the imperial system and the International System of Units. It is important to strictly follow the instructions for authors while preparing the manuscript to avoid an immediate rejection of the manuscript following the submission, without sending it further for peer review.

1.3 What Is Writing Well?

Although everyone has the right to write an aquaculture manuscript, not everybody knows what writing in aquaculture research and science is really about and how to write in the way that increases the chances of the manuscript getting into print. That is probably why most submitted manuscripts are rejected, and many experienced aquaculture scientists and educators complain about the quality of the papers that are published and made available in the public domain by "predatory journals", which can be defined as paid journals that care more about making money than publishing papers that contribute to the advancement of knowledge. Although the authors cannot always be blamed for publishing low-quality research, they can be blamed for low-quality writing. I strongly believe that some authors just don't invest enough time in learning how to write well and take advantage of the availability of predatory publishers to push out low-quality papers, which have been rejected by "serious" journals.

1.4 What This Book Will Do for You

Although there is nothing such as best writing, there is a way of editing and presenting an aquaculture manuscript that is considered as good practice by reviewers and editors and makes your research easy to understand. This book is written with the intention of improving the way less-experienced aquaculture scientists and students approach writing or changing the "bad" writing habits of seasoned researchers who struggle to get into top-flight journals. This book argues that a good presentation of an aquaculture manuscript, using appropriate writing tools and techniques, will increase the chances for getting accepted and published. wwwguarantee that the paper will be accepted, as the originality of the research, a sound methodology and a suitable choice of receiving journal, among other factors, are also important determinants of the acceptance of a manuscript for publication. Taking this into account, this book will provide the reader with the elements that are needed to appropriately plan and execute the writing of aquaculture manuscripts, including peer-reviewed manuscripts, scientific research reports, working papers, abstracts for aquaculture conferences, manuscripts for conference proceedings and technical manuscripts for aquaculture magazines, with a special emphasis on writing peer-reviewed aquaculture manuscripts. The reader will also receive insights on how to effectively submit aquaculture manuscripts to academic journals, review manuscripts for an academic aquaculture journals, respond to reviewer's comments and manage a manuscript writing project.

2

Principles of Scientific Writing

Scientific writing is both an art and a science.

Scientific writing is both an art and a science. It is an art because it tells the story of a set of scientific aquaculture activities and presents the findings in a way that keeps the reader awake and willing to keep reading. It is a science because the writing principles are rigorous and non-flexible and specific parts of the aquaculture manuscript such as the introduction, the materials and methods and the discussion cannot be skipped or ignored in any original research article. In order to clearly tell the story of the research that led to aquaculture findings, the aquaculture manuscript should be written using the available best practices to pragmatically frame the manuscript and avoid loopholes.

2.1 How to Frame and Focus a Paper

Each manuscript should focus on advancing knowledge on one subject, by providing solutions or tentative solutions to the research problem addressed. It is important to clearly define the research problem in a way that presents the extent of the problem and describes previous approaches to tackle the problem. For instance, if the authors want to talk about a new natural substances to treat a fish disease, the author could start with an opposite approach to disease treatment

DOI: 10.1201/9780429322648-2

in aquaculture, that is the use of chemical and medicine, then present the limitation of the use of medicines and then gradually move toward the use of natural products and their beneficial effect. As such, a way to frame a paper is to first introduce the current way of dealing with the problem and then introduce the new idea that will either be confirmed or disproved through the research that will be described in the subsequent sections of the manuscript. Such a focus on the main subject of the manuscript could be illustrated in the way sentences are written. Using the same example previously, every sentence that has the "natural substance" as the subject of the sentence orientates the focus of the reader toward the subject. For instance, in the sentence "*The natural substances* used in aquaculture include...", the subject of the sentence is clearly "the natural substances" and this captivates the attention of the reader. In contrast, in the example "*Aquaculture* uses several natural substances, including...", the subject is *aquaculture,* and this diverts the attention of the reader from the main subject of the paper which is the *natural substances.*

2.2 Loopholes That One Has to Consider When Writing

Writing a scientific manuscript is not generally about an author sharing knowledge. It is rather about sharing the results of a specific scientific experiment, in the hope that these results will improve the readers' perception on the subject. In other words, the intention of the authors is to make sure that the readers find the content of the scientific document understandable and of interest, so that they can use the information and cite the paper as it adds to their understanding of the subject. Authors should thus avoid putting themselves in a commanding position, where they impose a certain view on readers, but rather nuance their writing and orientate the reader, giving the opportunity to the readers to draw their own conclusion on the topic. This does not mean that the authors should not demonstrate their control over the research or share their perception of the realistic conclusion in the manuscript, but it means the conclusions and recommendations made by the authors should be motivated by the results of the experiment and presented as suggestions and definitely not as absolute opinion. In addition, each piece of research usually only contributes a tiny part to the larger research subject, which in turn represents a fraction of a research topic or a scientific discipline. As such, the authors should be very careful while drawing conclusions on the work they perform. Some authors pretend that their studies *demonstrate for the first time ever that…*, or *is the first study on…* or *now clearly shows that* or *is absolute evidence that…*, but then they suggest other studies that need to be conducted in order to either confirm or better understand their findings. It is therefore important that the authors keep a humble and non-pretentious tone while presenting the relevance of their findings to the aquaculture community.

2.3 Is English Grammar (Including Tenses) a Problem?

Although it is important to master the grammar and vocabulary of any language prior to writing in that language, the way the words are composed to form sentences in any scientific document is straightforward, as precision and concision are important syntactic considerations in science writing. Thus, if any author masters the scientific grammar of a specific language and knows how to write scientific articles in that language, then they will not have much difficulty when it comes to writing in English, provided that they are fluent in English. In addition, many writers consider that the structural rules governing the way the words are composed to form sentences in English are quite straightforward, which might be the reason why English has become a widespread language in the global scientific community. So there is no reason for non-native English speakers, who are fairly fluent in English, to consider English grammar to be the main excuse for their poor writing skills in English. Scientific writing style is universal, irrespective of the language.

3

Ethics in Scientific Writing

Science is rigid while research is flexible.

In aquaculture research in the modern world, like in any other creative discipline, ethics play a crucial role, governing such issues as the integrity of the scientist, the originality of the research, plagiarism and the management of animal subjects.

3.1 Being a Scientist

A scientist is someone who knows how to apply a scientific approach to experimentation in order to draw scientifically sound conclusions from the experimentation. There is no special achievement that strictly leads to the bearing of the professional title of "scientist". In other words, everyone can pretend to be a scientist, and no one has the right to refute this title to anyone else. However, advanced education, the practice of science and science productivity (articles, patents, etc.) are common attributes that confer the recognition as "scientist" to a practitioner.

Science is a rigorous endeavor that requires scientists to be open to exploring new ideas while applying rigid scientific approaches that are based on a high degree of ethics. As such, scientific ethics are characterized by values such as integrity, responsibility and accountability and respect. "Integrity" refers to the fact that the scientist should avoid

DOI: 10.1201/9780429322648-3

immoral practices such as plagiarism, misappropriation of others' ideas and findings, dishonest reporting of facts and findings and non-recognition of the contributions of others in a research activity. "Responsibility" refers to the use of time and material resources (including the budget) in a way that advances the work for which they were assigned and not to support personal interests. "Accountability" refers to the fact that the authors are responsible for the information included in the manuscript to the point that, in case an error is discovered even after the manuscript has been published, it is the authors' responsibility to contact the publisher immediately to submit an erratum in order to rectify the error. "Respect" refers to considering everyone else in a research team or organization with respect and showing appreciation to the staff, colleagues, funders and partners involved in the research process. The lack of these values in a scientist could be hidden in the short run, but they are always recognizable in the long run, and could lead to distrust toward the scientist and his/her work and finally tarnish or terminate his/her career. No one wants to collaborate with a scientist who lacks scientific ethics. Many scientists have been fired from research organizations for their lack of ethics, although it is not a subject people like to talk about.

3.2 Originality of the Research

For any research to be published and to be used to produce any patent or to be exploited by the end users, it needs to be original, meaning that the study is not redundant, that nobody has done the same work anywhere else in the world before and the results must not be fabricated or misreported by the authors. A way to avoid redundancy in research is to conduct an extensive, thorough literature review on the subject covered by the research. Another way to guarantee the originality of research is an internal peer review of research protocols and concept notes, at the department or organization level, prior to the execution of the research. It is unacceptable to deliberately use someone else's idea, materials or methods to advance one's research, without giving credit to the initial authors of the idea, materials or methods. Hence, the proper citation of the original source of an idea is extremely important in science, as it gives the scientist the opportunity to respect the authenticity of the previous work, while building on that previous idea to highlight the additional value that his/her own work will contribute to the idea. Giving credit to author of other papers does not guarantee that plagiarism is avoided, as there are wrong and right ways of quoting and paraphrasing a published text or method.

3.3 Citing a Source

Scientific writing involves using published information to present a research problem, to describe the materials and methods used and to discuss the results obtained from original research. Specifically, most of the materials used in aquaculture research are developed by engineers while the methods are developed by biologists. It is important to clearly describe the materials used such as equipment (name, brand, country of provenance) and methods in which these materials are used. If the methods have already been published in another peer-reviewed paper, it is not necessary to describe it again in an aquaculture manuscript. A reference to the other paper could suffice. However, if the author did some modifications to the methods described by another author, details of the modification should be provided. Nevertheless, although borrowing ideas from other authors is part of the writing process, not properly citing the source of the borrowed information is considered plagiarism.

3.3.1 Plagiarism

Plagiarism is considered as one of the most unethical practices in science, as it is not fair for an author to take credit for other authors' efforts and creativity. Detecting plagiarism in manuscripts is not an easy task for reviewers, editors or publishers of scientific journals, especially considering that there is a myriad of scientific journals published in several different languages. However, when plagiarism is detected in a scientific publication, the consequences are usually disastrous and might lead to disgrace and firing of the plagiarizer, and, in extreme cases, a lawsuit. Furthermore, plagiarism might not only destroy the plagiarizer's reputation and career but can also tarnish the image of the research institution or university where the plagiarizer is affiliated. Citing a reference in a scientific text does not necessarily prevent plagiarism, as there are specific rules that guide the proper use of citations, whether the author is copying and pasting, paraphrasing or translating someone else's published idea.

3.3.2 Rule for Properly Copying and Pasting Published Information

When an author decides to copy and paste another author's idea, word by word, the borrowed text must be put into quotation marks and information (usually the name and year of publication) on the original source provided. For instance, let's consider that an author named Peugoue published the following statement in a book in 2010: "fish is the most nutritious food". Here are two possible ways of properly copying and pasting this statement in a scientific text:

1. According to Peugoue (2010), "fish is the most nutritious food",
2. "Fish is the most nutritious food" (Peugoue 2010).

3.3.3 Rule for Properly Paraphrasing Published Information

Paraphrasing means presenting someone's statement with different words and language style, without changing the original meaning of the statement. Any case of paraphrasing in a scientific text must be accompanied with information (usually name and year of publication) of the original author of the idea. Using the same example previously, here are two possible ways of properly paraphrasing Peugoue's statement:

1. Among all the foodstuffs that exist, fish has the highest nutritional value (Peugoue 2010),
2. Peugoue (2010) previously asserted that fish occupies the first position among the most nutritious foodstuffs.

3.3.4 Rule for Properly Translating Published Information

Although nowadays it is well accepted that the language of science is English, several highly regarded scientific journals are still published in other languages. However, directly translating someone's idea from an initial language to another language is also plagiarism. Hence, when a word-by-word translation is performed on a borrowed idea, the source of this idea must be cited with quotation marks and original author's information provided as per the rule of copying and pasting presented previously. In the situation where a borrowed idea is translated and paraphrased, the source of the original idea must be properly cited following the rule of paraphrasing described earlier.

3.3.5 Self-Plagiarism

Plagiarism also includes "self-plagiarism", which occurs when an author reuses his/her previously published idea in a new publication without properly self-citing the initial source (Masic 2012). Therefore, the same rules for properly citing references also apply to citing one's own previous work.

As an Editor of an academic aquaculture journal, the most frequent cases of plagiarism I have encountered in aquaculture manuscripts are "self-plagiarism". It is also common to see cases of copying and pasting someone else's idea with the initial author's information (name and year of publication) provided but without quotation marks to highlight that the idea was borrowed word by word from elsewhere. Therefore, inexperienced aquaculture authors should not only keep on citing the sources of their borrowed information but also make sure they properly cite this information in the text of their scientific manuscripts. It is important to know that the publisher uses specialized software to detect obvious cases of plagiarisms during the production of an accepted manuscript. Should plagiarism be detected, the accepted manuscript is either sent back to the authors, for proper citation, or is totally banned from publication.

3.4 Animal Ethics

Animal ethics is the component of the moral principles that examine the relationship between the humans and animals, with a focus on how living animals have to be handled and treated in captivity by humans. It is well established that for an animal to express its full growth and development potentials, it has to be treated with dignity and without receiving any needless non-experimental pain or distress. In science, animal ethics is governed by institutional, local, national or regional animal ethics boards, whose mission is to receive, analyze and grant approvals for animal ethics to applications submitted by scientists conducting research that involves living animals. Complying with the animal ethics rules and regulations applicable to a specific context is a prerequisite for the submission and publication of manuscript in most peer-reviewed aquaculture journals.

In practice, animal ethics guarantees that any aquaculture research conducted with living animals such as fish, crustaceans, frogs and crocodiles complies with the "three R's", which are replacement, reduction and refinement (Russell and Burch 1959). "Replacement" means that the investigators should try as much as possible to avoid using living animals and replace them with other nonliving parts of animals such as cells for instance. "Reduction" means that in case the use of living animals is absolutely necessary, then the investigators should limit the number of animals used to the strict minimum necessary to achieve the research goals. "Refinement" means that the investigators should create the best rearing and caring conditions as well as appropriate management and sampling procedures to keep the possible pain and suffering endured by the animal to the minimum. It is important that the investigators think about animal ethics early in the research process, during the preparation of research protocols, to ensure that approval by animal ethics authorities is made easy. In many research organizations, research involving living animals cannot effectively start before the animal ethics approval is obtained from a recognized animal ethics board. In some countries, the investigator may have to pay a fee on receiving the approval of their application, prior to receiving the certificate for animal ethics approval. In many peer-reviewed journals, a manuscript resulting from a study that involved the use of living animals cannot be submitted/accepted if the investigators do not provide a proof of animal ethics approval obtained from a recognized animal ethics board.

4

Writing a Peer-Reviewed Original Article

Publishing your manuscript in a peer-reviewed journal is a good way to validate your research results and findings.

This chapter is adapted from my previous publication, Yossa (2014). The global average acceptance rate of manuscripts submitted for publication in scholarly journal is 35%–40% (Aarssen et al. 2008; Björk 2019), and important among the reasons for rejection is poor presentation. Regardless of language, well-organized and clearly articulated papers will be considered more seriously and used more frequently than those difficult to read. General guidance on issues to consider in producing a scientific research article is available (Cargill & O'Connor 2013; Carpenter 2001; Day 1994; Elefteriades 2002; Hengl & Gould 2002; Jennings et al. 2012; Shubrook et al. 2010), and these general principles also apply to aquaculture research publication. Here, we focus on the aspects of specific relevance to scientific writing in aquaculture research to help students and inexperienced aquaculture scientists improve the acceptance rate of their submitted manuscripts. We begin with a review of basic concepts in the preparation of an experiment, such as defining the problem and preparing the execution of the study, and continue with a

DOI: 10.1201/9780429322648-4

thorough description of the major elements included in a scientific manuscript that will be submitted to a peer-reviewed academic journal for publication.

4.1 Defining the Problem

Many research results submitted for publication are rejected because they have already been done by others. Scientific progress is cumulative, building bit by bit from simple principles to gradually understand complex phenomena. As Isaac Newton wrote in 1676, "If I have seen further it is by standing on the shoulders of giants." Each new generation benefits from the lessons of the past but only if they know what has already been done. Repeating research to test established fact can sometimes be useful in situations where general knowledge needs to be adapted to local conditions, but this kind of research is best placed in local extension bulletins or trade magazines and is seldom accepted by international journals. The thorough literature review is thus an essential part of the problem definition and the choice of methods. To complete a literature review properly, it is absolutely essential for any working scientist to be up to date in his or her field of expertise.

4.2 The Research Plan

A confusing manuscript often starts with a confusing research plan, also called research protocol. Before doing anything else, the researcher should carefully define the question being addressed (e.g., the hypothesis) and the methods and statistical model that will be used to test the hypothesis. Sometimes, methods used by others can simply be repeated. New technology or approaches to problem-solving, however, can present opportunities to study phenomena that were previously impenetrable. Aquaculture scientists are well advised to build bridges and collaborations with partners in other fields and be open-minded to new ideas that can produce innovative research. Research plans are not only helpful in ensuring that common mistakes in aquaculture research (e.g., using individual fish as replications when tanks or ponds are the actual unit upon which a treatment is applied) are avoided but they are also often important components of grant proposals. A well-structured research plan should include detailed information on the following:

- A clear statement of the hypothesis
- The start date and the duration of the experiment
- The experimental (statistical) design, including layout of treatments and replications
- Treatments and sampling protocols
- Laboratory, computer and other equipment and materials
- Expected results
- List of participants to the research and their roles.

It is always a good idea to request other members of the research team and/or colleagues from elsewhere to review the research plan, in order to help identify fatal flaws that might waste time and money and preclude publication. The research plan should be made available to the entire research team, and salient features (e.g., sampling protocols, feeding levels, water quality testing) should be posted. Whenever there are any questions about how to proceed, one can refer back to the research plan.

4.3 Preparing a Manuscript

Upon completion of the experiment and analysis of the data, the entire research team and co-authors/investigators should be made aware of the findings, then the responsibility of each person with regard to the writing of the resulting manuscript defined and the order of authorship agreed. Order of authorship is generally based on one's participation to the following steps:

- Definition of the problem
- Development of the research plan
- Execution of the research plan
- Statistical analysis
- Manuscript production (writing)

Each of these steps is often led by different people within a team. How the workload is managed varies according to country and culture, but to be consistent with international scientific practice, the usual one to two people who are mostly responsible for the research and can answer detailed questions should be the first authors. The corresponding author, who has prepared the manuscript and managed the review process and who will be easily located by people with questions, is the person who can best answer questions. For example, a major professor may be the corresponding author in cases where a student does not know where she/he will be in coming years. People who have political authority, but no special ownership of the paper, are typically acknowledged in a separate section but are not listed as authors.

Depending on the weight of the information that it contains, a manuscript submitted for publication as an original article can be accepted for publication as it is or as a short communication/note. In the original article, the data contained in the manuscript cover the entire subject, while in the short communication/note, the data are just a part of the subject or a preliminary work to a further more elaborated study. As such, the final decision on the type of article to be published will be made by the editors, in agreement with the authors.

4.4 Organization and Formatting

Each scientific journal has its own style and format, and manuscript formatting must comply with the "instructions/guidelines for authors" of the selected journal. This includes such things as format for margins, font, spacing, line numbering, reference style, graphics and tables. While editors vary in the degree to which they insist on formatting, most manuscript submissions these days are done through an unforgiving online submission system, so it is generally a waste of everyone's time to send in manuscripts that do not conform to the journal's style. They will simply be returned to the corresponding author for compliance.

By and large, scientific manuscripts are comprised of eight sections as follows:

- Cover page
- Abstract
- Introduction
- Materials and Methods
- Results

- Discussion
- Acknowledgments
- References

Sometimes, Results and Discussion are combined. Sometimes, there is a separate Conclusions section after the Discussion. Although it is difficult to propose a figure that is applicable to all manuscripts, the approximate length of each of the main sections of a manuscript (in percent) should be: Cover Page/Abstract/Acknowledgments (10%), Introduction (10%), Material and Methods (20%), Results (25%), Discussion (25%) and References (10%).

One key aspect of proper manuscript organization that is often overlooked is the order in which ideas are presented. If an experiment has subcomponents, these should be always treated in the same order in the Abstract, the Introduction, the Materials and Methods, the Results, and the Discussion. Keeping things in order this way helps readers find what they are looking for, without having to wade through the entire article.

4.4.1 Cover Page

The cover or title page lists the title of the manuscript, the names, affiliations and contact details of the authors and indicates the corresponding author. In some cases, the authors are requested to provide a suggested running title that will appear on the top of each page of the manuscript when published. The running title, which usually comprises a maximum of 50 characters with spaces, should capture in a few words the gist of the article that helps the readers know at which article they are looking. For example, an article entitled: "Dietary Protein Requirements of Cameroon Major Carp Fingerlings Grown to Market Size in Ponds in Bafang" might have a running title of "Bafang pond diet for carp".

Also common on the title page are five to seven keywords that can be used in literature or online searches to find the article. To this end, the words already in the title should not be repeated as keywords, as any search will already find them. The keywords should attempt to capture the broader context in which the research was done so that researchers in related fields can find the article. For the previously mentioned example, keywords might include the names of specific Cameroon carps (*Cyprinus carpio*), the region in which the results should be applicable (Central Africa, tropics), the specific habitat in which the system operates (grassfield, highlands, Bamiléké), the reason for growing carp in Cameroon (food security, rural livelihoods) or other aspects of the culture system (juvenile nutrition).

The title should be explicit and as short as possible, capturing the main ideas but not overdoing it. There are several types of titles:

- Interrogative: "Does Fry Grading prior to Stocking Improve the Production of Channel Catfish, *Ictalurus punctatus*, in Recirculating Aquaculture Systems?"
- Affirmative: "Fry Grading prior to Stocking Improves the Production of Channel Catfish, *Ictalurus punctatus*, in Recirculating Aquaculture Systems"
- Descriptive: "Effects of Fry Grading prior to Stocking on Fish Production of Channel Catfish, *Ictalurus punctatus*, in Recirculating Aquaculture Systems".

In any case, the title generally includes information on the treatment (e.g., "grading"), the species of fish (e.g., "common and scientific name"), life stage of the fish species studied (e.g., "fry") and the production system (e.g., "recirculating aquaculture systems").

4.4.2 Abstract

Once a reader has found an article with an interesting title, she/he usually first reads the abstract. For articles in online publications that charge for their content, often the abstract is the only part most people will read. The abstract is generally written in one short (200 words is a long abstract) paragraph. Normal abstracts should include only salient features of the study, including objectives, treatments, main results and conclusions/recommendations. First in the abstract should be a statement of the objective of the study, for example (from Yossa et al. 2011):

> A study was conducted to investigate the effects of dietary avidin on growth, survival, food conversion, biotin status and gene expression in juvenile zebrafish, *Danio rerio*.

The second part of the abstract should describe the treatments and the main methodology used. For example (from Yossa et al. 2014):

> Six isonitrogenous and isocaloric purified diets containing 0.031 (biotin-unsupplemented diet), 0.061, 0.263, 0.514, 1.741 and 2.640 mg biotin kg^{-1} diet were fed in triplicate tanks for a total of 12 weeks to juvenile zebrafish (initial mean body mass 0.13±0.001 g).

Results reported in the abstract are only those directly relevant to the title and objectives of the study, for example (from Yossa et al. 2015):

> Following the feeding period, males fed biotinsufficient diet exhibited higher gonado-somatic index, sperm density, sperm motility and sperm viability than those fed biotin-unsufficient diet ($P<0.05$). In the presence of biotin-sufficient males, biotin-sufficient females spawned more eggs (222.2 eggs) than biotin-deficient females (18.8 eggs) ($P<0.05$). The same pattern was

> observed with biotin-deficient males (7.6 vs. 1.8 eggs) ($P<0.05$). Biotin-sufficient males generated a higher percentage of fertilized eggs (90% vs. 42%), hatching rate (62% vs. 27%), larvae survival (98% vs. 37%) and larvae length at 7 days post fertilization (4.4 mm vs. 4.2 mm) than biotin-deficient males ($P<0.05$).

The last sentence of the abstract is the conclusion (and recommendation) of the study, which is also the take-home message (from Yossa et al. 2018):

> The blood smear method is therefore a valid, simple, and less expensive alternative to flow cytometry for the assessment of ploidy level in Arctic charr.

It is not appropriate in the abstract to include the background of the study or discussion of the results. Citations and abbreviations are also not included.

4.4.3 Introduction

The Introduction defines technical terms and expressions, presents the research problem, summarizes the state of knowledge and previous attempts to solve the problem, explains what this new study will do (objectives) and why it is relevant and clearly states the hypothesis to be tested. In this section of the manuscript, the choice of fish species and production system for the study should be explained relative to the global and/or local production of the fish, problems encountered in the production cycle, etc. It should be pointed out that the introduction of a scientific journal article is not the same as the literature review in a thesis or dissertation. It is only in the rare case where a problem is extraordinarily complex that the Introduction exceeds one page in length. Also, for papers submitted to aquaculture journals, it is not necessary to state the importance of aquaculture or describe how diseases or nutrition are important. Any reader of an aquaculture journal will already be familiar with these issues. Keep in mind that the order in which issues are raised in the Introduction should be maintained throughout the paper.

4.4.4 Materials and Methods

The Materials and Methods section is like a cookbook. It can be considered as step-by-step instructions to anyone interested in repeating the study, either to confirm or challenge the findings. This is an essential part of the scientific process, and researchers should make every effort to include the details necessary to exactly replicate the study. In particular, any irregularities (e.g., loss of a replication) should be carefully explained. The materials and methods derive directly from the research plan. The exact order in which the methods are described should be roughly chronological. For example, source of the fish, experimental units, replications, feeding, water quality, sampling, harvest, laboratory procedures, calculations and statistical analysis. As with the Introduction, subjects treated in the Materials and Methods section should be maintained in the order in which they are first mentioned. Common subheadings within the Materials and Methods section often include Fish, System (or Facility), Experimental Design, Feed and Feeding, Sample Collection and Analytical Procedures, Measurements/Calculations and Statistical Methods.

4.4.4.1 Fish

In this subsection, information on the source, species, strain and initial size of fish and name of the fish supplier is provided. Any treatment prior to the start of the experiment (e.g., quarantine, type of feed provided, feeding level and frequency, acclimation period, grading) should be mentioned.

4.4.4.2 Facility

The production units (e.g., pond, aquarium, tank) should be described in terms of construction material (e.g., earthen, concrete, plastic), color, dimensions, actual volume (often different than total volume) and, if relevant, arrangement. The latter can be especially important in cases where interaction between units might be significant, such as ponds in a series or tanks in a row where one receives more light or disturbance than others. Depending on the nature of the study, light intensity and photoperiod can be important. Also in this subsection is a description of the water supply (well, surface water, tap), water exchange (recirculating, flow-through, static) and range of physicochemical parameters of the water (temperature, pH, salinity, hardness, conductivity, nitrite, nitrite, ammonia, etc.), and all the actions taken to maintain desirable water quality and quantity should be presented.

4.4.4.3 Experimental Design

In this subsection, the treatments, the experimental units, the sampling units, the replications and the experimental plan (e.g., completely randomized, complete blocks, factorial) are presented. To save space in the manuscript, treatments are often given short labels (e.g., T1, T2 or High Input, Low Input). This should be done in such a way that is intuitive to readers.

4.4.4.4 Feed and Feeding

In this subsection, the type of feed (purified, semipurified, practical), form of feed (moist pellet, dry pellet, powder, etc.), feeding level (restricted, demand, satiation), feeding frequency (number of feedings per day and explanation of how the total ration is divided) and feeding method (hand feeding or automatic feeding) are presented. Unfed treatments are generally a thing of the past in aquaculture research. We all know now that unfed fish do not grow! Practical information on methods for collecting uneaten feed and the calculation of the feed conversion ratio should also be reported.

The proximate analysis of the feed(s) should be presented, usually as a table, and any irregularities noted. Usually, proximate analysis includes crude protein, crude lipid, crude ash, crude fiber, total carbohydrate and gross energy. In some cases (e.g., nutrition studies), information on the fatty acids and amino acids profiles of some ingredients and the diets can be provided.

4.4.4.5 Sample Collection and Analytical Procedures

In this subsection, information on the periodicity of sampling (start of the experiment, end of the experiment, every week, etc.), the type of sample (whole fish, liver, kidney, etc.), sample processing for storage and the storage conditions (temperature, duration, etc.) is presented. References for standard methods should only be cited, for example: "The method used to count the cells was previously described in Ngaha (2004)". However, details on any modifications or nonstandard methods should be thoroughly described, for example: "the cell motility was analyzed using the method described by Djiethieu (2006); however, instead of using the heating temperature of 26°C as in Djiethieu (2006), the water temperature was maintained at 24°C".

4.4.4.6 Measurements and Calculations

The formulae used to calculate the study parameters (with references) are provided in this subsection. For example: "feed conversion ratio, FCR (g/g) = [(Quantity of dry feed distributed (g)) × Dry matter content of feed]/[final wet mass (g) – initial wet mass (g)]".

4.4.4.7 Statistical Methods

This section is modified from Yossa and Verdegem (2015). Statistical methods (*T*-test, ANOVA, ANCOVA, regression, etc.), significance level ($p<0.05$ or $p<0.01$), multiple comparison test (Duncan's multiple range test, Tukey's test, etc.) and name and version of the software used to perform statistical analysis on the data must be mentioned. The scientific interpretation of results is governed by statistics, which in turn support the decision made on any results. Hence, the use of the right statistical method is crucial for the accurate interpretation of the results and for drawing the right conclusions. The discussions around the statistics (experimental design, treatments, number of replicates per treatment, number of sampling, significance level and statistical software) have to happen during the design of the experiment, prior to its execution. In this manner, all the resources available for the execution of the experiment will be efficiently used, and the expected output of the experiments will be anticipated. This planning part of the statistics is very important for the success of an experiment, as once the experiment is executed, it is impossible to improve the design, and any attempt could be unethical or fraudulent. There is a book titled *Statistics for Aquaculture* (Bhujel 2009) that was published on this subject. Readers are encouraged to read this book, which also contains questions and practical exercises.

More recently, I also co-authored a peer-reviewed article on the "Misuse of multiple comparison tests and underuse of contrast procedures in aquaculture publications" (Yossa and Verdegem 2015). This paper critically analyzed the statistical methods used in articles published in ten selected international peer-reviewed aquaculture journals in the year 2013. This analysis showed that in none of the studies in which the independent variable was qualitative with a structure, were the data analyzed using orthogonal contrast procedure. Also, the data of only 34% of the studies in which the independent variable was quantitative have been analyzed using polynomial contrast (regression), whereas the data of only 13% of studies with a factorial design have been analyzed using contrast procedure. The conclusion of the paper was that more attention should be paid on publishing only studies that used appropriate statistical procedures, which conform to the nature of the independent variables of interest. The readers are encouraged to read the full article. However, we are going to reiterate part of this paper in the next section, in order to emphasize the necessity to make the distinction between the multiple comparison tests and the contrast procedures as they apply to aquaculture research.

Characterizing the independent variable

For every finfish and shellfish belonging to the 600 aquatic species currently raised in captivity worldwide, aquaculture research essentially focuses on investigating the range of water physicochemical parameters (temperature, pH, hardness, ammonia, dissolved oxygen, etc.), nutrient requirement and substitution levels (macronutrients such as proteins, lipid and carbohydrates and micronutrients such as vitamins and minerals), feeding (feeding rate, feeding frequency, etc.), use of therapeutants and production system and technology (light intensity levels, photoperiod, animal density, etc.). In other words, in aquaculture as in other sectors of agricultural sciences, most of the independent variables (experimental variables) are qualitative (with a structure), quantitative or factorial combinations. The use of qualitative variable with a structure and quantitative experimental variables, alone or in combination, allows making "planned comparisons" (Steel et al. 1997).

A qualitative independent variable is a variable that has unquantifiable, nominal variants (levels), which represent the different categories included in this variable such as the presence or absence of a character or the fish gender (male or female). The **structure** in a qualitative independent variable refers to the existence of a relation between the different variants of this variable, in a way that suggests that some variants can be grouped together and then compared to other groups of variants (Steel et al. 1997). For example, assuming that the objective of an aquaculture experiment is to determine the dietary protein optimizing fish growth, there might be four protein types (treatments) tested: soybean meal, corn gluten meal, fish meal and beef blood meal. There is a structure among these protein types as the two plant proteins (soybean meal and corn gluten meal) can be grouped and compared with a group including the animal proteins (fish meal and beef blood meal). In addition, soybean meal can be compared with corn gluten meal and fish meal can be compared with beef blood meal. Another example is an aquaculture experiment aiming to identify the plant protein that when substituting 10% of fish meal in a salmon diet results in the best growth. Four experimental plant proteins (treatments) might be tested: corn gluten meal, soybean meal, black eye pea meal and canola meal. There is no relevant relationship among the experimental plant proteins that can be considered in order to classify them into definable treatment groups; therefore, there is no structure in the independent variable of this experiment.

A quantitative independent variable is a variable with levels that are measurable quantities that are expressed numerically such as the water temperatures, dietary protein levels or fish densities per tank. The levels of a quantitative independent variable are maintained unchanged throughout the experimental period. An example is an experiment in which the independent variable is the water temperature, with five levels, 20°C, 22°C, 26°C, 28°C and 30°C.

In a study with one independent variable (unifactorial experiment), each variant of this variable represents a treatment, and the total number of treatments (k) equals the total number of variants of the independent variable. In a study with two or more independent variables, also called a factorial or multifactorial experiment, the total number of treatments (k) represents all the possible combinations of the two or more independent variables (Petersen 1977; Steel et al. 1997). For instance, a study with the independent variables (factors) water temperature with two levels (20°C and 30°C) and dietary protein source with three levels (soybean meal, gluten meal and black eye pea meal) is a 2×3 factorial experiment, with 2×3=6 treatments (Table 4.1).

Several studies have pointed out the extensive misuse of pairwise multiple comparison tests in agricultural sciences (Chew 1976; Gates 1991; Gill 1973; Little 1981; Madden et al. 1982; Petersen 1977). However, the examination of the use of statistics in aquaculture science has received little attention. In order to draw the attention of researchers on the importance of good statistical planning and proper data analysis, we are going to review some basic and applied

TABLE 4.1

Example of the Different Treatments Obtained by Combining the Levels of the Two Independent Variables (Protein Sources and Water Temperatures) in a Factorial Experiment

		Proteins Sources		
		Soybean Meal (SB)	**Corn Gluten Meal (CG)**	**Black Eye Pea Meal (BEP)**
Water temperatures (°C)	20°C	SB20[a]	CG20[a]	BEP20[a]
	30°C	SB30[a]	CG30[a]	BEP30[a]

[a] SB20 refers to soybean meal fed to fish reared at 20°C, SB30 refers to soybean meal fed to fish reared at 30°C, CG20 refers to corn gluten meal fed to fish reared at 20°C, CG30 refers to corn gluten meal fed to fish reared at 30°C, BEP20 refers to black eye pea meal fed to fish reared at 20°C and BEP30 refers to black eye pea meal fed to fish reared at 30°C.

concepts in relation to the planning and execution of data analysis (and interpretation).

Multiple comparison tests versus contrast procedures
In an experiment with one independent variable that has more than three levels, and assuming that the independence, normality and homogeneity of the residuals are satisfied, analysis of variance (ANOVA) is used to test the null hypothesis: "there is no significant difference between the treatment means". In an experiment with a combination of two or more independent variables (factorial experiment), the null hypothesis is tested using (multi)factorial ANOVA (two-way ANOVA, three-way ANOVA, etc.). When there is a significant difference among the treatment means, the null hypothesis is rejected and a multiple comparison test, an orthogonal contrast procedure or a polynomial contrast procedure is applied to separate treatment means or to analyze the relationships that might exist among the treatment means. The choice of one of these statistical procedures depends on the nature and structure of independent variables (qualitative with or without structure, quantitative or factorial combinations).

Multiple comparison tests are statistical methods that are intended to compare each factor level mean with every other factor level mean assuming that there is no definable structure among the factor levels (Gates 1991; Montgomery 1997; Petersen 1977; Saville 1990; Steel et al. 1997). In this manner, a pairwise multiple comparison compares factor level effects on the response variable. For instance, in the above experiment with the four experimental plant proteins, the use of a multiple comparison test is relevant, as no definable structure exists among the experimental plant proteins (treatments). Examples of multiple comparison tests include Least Significant Difference (LSD), Duncan's Multiple Range, Waller and Duncan's test, Tukey's Honest Significant Difference (Tukey-HSD) and Bonferroni and Scheffé's procedures. The merit of each of these tests over the others has extensively been discussed (Montgomery 1997; Steel et al. 1997) and is beyond the scope of this paper. However, with qualitative independent factors showing a structure, statistical procedures exist which lead to more meaningful conclusions on treatment effects; furthermore, pairwise multiple comparison tests should not be used on quantitative main factors and factorial interaction terms as this leads to misinterpretation of research results and flawed conclusions (Gates 1991; Montgomery 1997; Olsen 2003; Petersen 1977; Rafter et al. 2002; Saville 1990; Steel et al. 1997).

In an experiment with a qualitative variable showing a structure, an adequate statistical procedure for comparing treatment means is **the orthogonal contrast procedure** (Davis 2010; Montgomery 1997; Steel et al. 1997). With this procedure, related treatment means or groups of treatment means are specifically compared to other treatment means, for a total of comparisons equal to the degrees of freedom. Each comparison is planned prior to the execution of the experiment and can represent an objective in the research protocol. For instance, in a fish nutrition experiment aiming at (1) selecting the best protein source for fish growth, (2) determining the best plant-based protein for fish growth and (3) determining the best animal-based protein source for fish growth, the independent variable is the fish diet formulation (qualitative variable). In this example, there is only one source of variation, the diet, and the four experimental treatments (k) that can be divided into two groups, one group composed of the two dietary plant-based protein sources (soya bean and corn gluten) and the other group composed of the two dietary animal-based protein sources (fish meal and beef blood meal). There is thus a clear relationship between these treatments, which means that the independent variable is structured. Assuming that the ANOVA will show that there is a significant difference among the treatment means, the following three (i.e., degree of freedom = k − 1 = 4 treatments − 1 = 3) contrasts can be planned prior to the execution of the experiment:

- Comparison of plant-based protein versus animal protein (experimental objective 1),
- Comparison of the two plant-based proteins (experimental objective 2),
- Comparison of the two animal proteins (experimental objective 3).

It is worth noting that the number of contrasts (3) is equal to the number of comparisons you would make with a post hoc test assuming an experiment with four treatments (diets), resulting in three post hoc comparisons, each with one degree of freedom. However, the contrast procedure offers the possibility of comparing groups of treatments, which increases the power of the contrasts over multiple comparisons; for instance, in this example, a multiple comparison would not specifically make the ith comparison.

On the other hand, the most relevant method recommended to analyze the means obtained from an experiment with a quantitative independent variable is **the polynomial contrast procedure** (Davis 2010; Montgomery 1997; Steel et al. 1997). Polynomial contrasts detect the trend of the relationship or regression that exists between an independent experimental variable and the response (dependent) variable. This relationship might be linear (linear regression), quadratic (quadratic regression or second-order polynomial) or cubic (cubic regression or third-order polynomial). For instance, in a dose response experiment aiming at determining the minimum vitamin level that leads to the highest growth in fish, the graded vitamin A levels (k) might be 0, 0.01, 0.05, 0.1, 0.2, 0.5, 0.7 and 1 mg/kg diet. The optimum vitamin A level is estimated by analyzing the

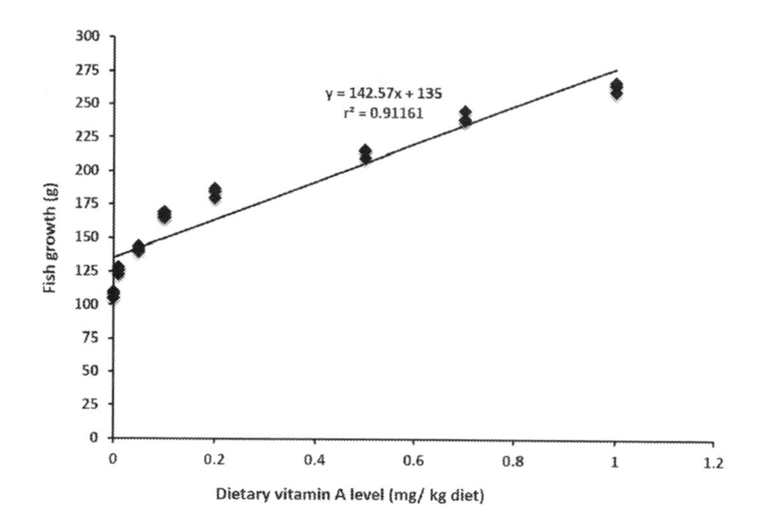

FIGURE 4.1
Example of the use of the polynomial contrast (regression) to analyze the data of an experiment in which there is a significant ($p<0.05$) linear regression between fish growth (dependent variable) and dietary vitamin A levels (independent quantitative variable).

regressions (linear, quadratic or cubic) between the growth and the vitamin levels. Let us assume that the regression is linear (Figure 4.1), that is, the curve function or regression equation is in the form: growth=X (vitamin A level)+Y, or growth=142.57 (vitamin A level)+135, where "X" (=142.57) is the slope and "Y" (=135) is the value of the growth when no dietary vitamin A is added to the diet; then the researcher will conclude that fish growth increases with an increase in the dietary vitamin A level.

Let us now assume that the regression is quadratic (Figure 4.2), that is, the curve function or regression equation is in the form: growth=X (vitamin A level)^2+Y (vitamin A level)+Z, or growth=–240.3 (vitamin A level)^2+294.2 (vitamin A level)+125.1, where "X" (=–240.3), "Y" (=294.2) and "Z" (=125.1) are the coefficients of the quadratic curve function, with "Z" (=125.1) being the value of the growth when

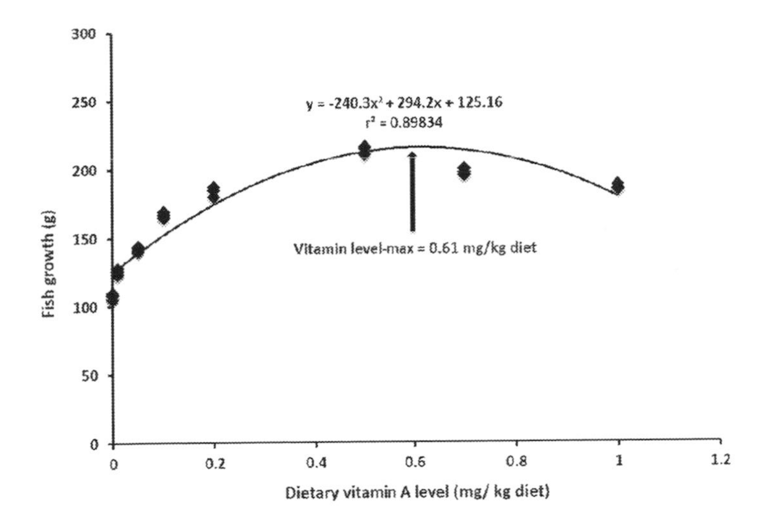

FIGURE 4.2
Example of the use of the polynomial contrast (regression) to estimate the dietary vitamin A level (0.061 mg/kg diet) that leads to the maximum fish growth, in an experiment in which there is a significant ($p<0.05$) quadratic relationship between fish growth (dependent variable) and dietary vitamin A levels (independent quantitative variable).

no dietary vitamin A is added to the diet; then the next step is to estimate the vitamin A level (vitamin A level max) that leads to the maximum growth of fish, that is, the inflection point which represents the point at which the curvature or concavity changes sign from minus to plus or from plus to minus on a curve. This inflection point is obtained through the equation in which the first derivate of the curve function is equal to zero; in the current example, d(growth)=0 $\rightarrow$ 2X (vitamin A level)+Y=0 $\rightarrow$ vitamin A level max=–Y/(2X)=–294.2/(2×–240.3)=0.612 mg vitamin A.

In a situation where there is a significant cubic regression (Figure 4.3a), the use of straight **broken-line regression procedure** has been suggested (Robbins et al. 1979, 2006). The straight broken-line regression procedure divides the data in two parts and assigns a linear regression to

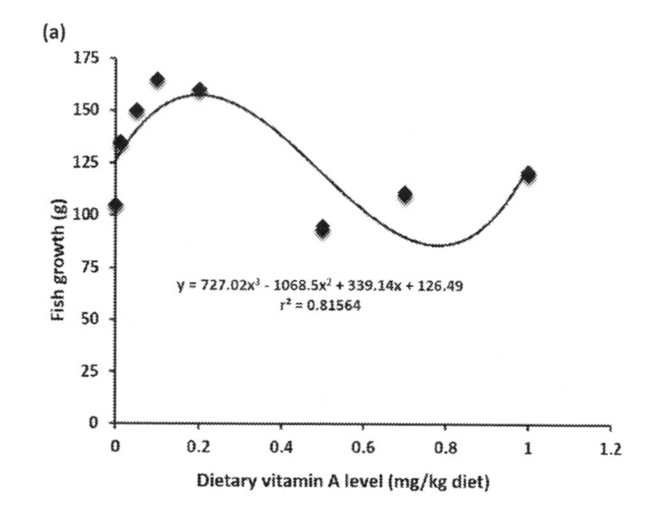

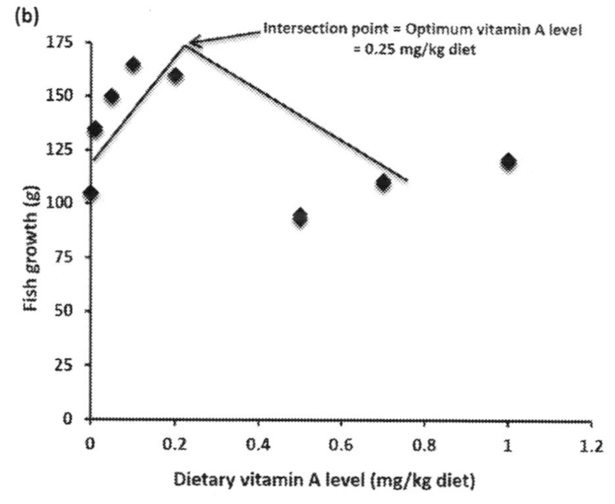

FIGURE 4.3
Example of the use of the polynomial contrast (regression) to analyze the data of an experiment in which there is a significant ($p<0.05$) cubic regression between fish growth (dependent variable) and dietary vitamin A levels (independent quantitative variable) (a); the broken-line regression procedure estimated the vitamin A level that leads to the maximum fish growth at 0.21 mg/kg diet (b).

each part (Figure 4.3b). The intersection point of the two linear regressions is the optimal level of the quantitative factor (Figure 4.3b). For more details on the broken-line regression procedure, the readers are encouraged to read Robbins et al. (2006).

Furthermore, the polynomial contrast procedure also provides the coefficient of determination "r^2", which measures the adjustment of the regression equation in regard to the points that illustrate the relationship between the independent and the dependent variable in a diagram (Figures 4.1–4.3); for instance, a coefficient of determination of 0.91 ($r^2=0.91$; Figure 4.1) indicates that the current regression equation explains 91% of the variation observed in the dependent variable (growth), as a result of changes in the independent variable (vitamin A); this information cannot effectively be provided with the multiple comparison procedures (Petersen 1977).

In practice, the first step in applying the polynomial contrast procedure is to check whether or not the cubic regression is significant. If the cubic regression is significant, then the quadratic and linear regressions are not considered; otherwise, the significance of the quadratic regression is examined. If the quadratic regression is significant, the linear regression is not considered; otherwise, the significance of the linear regression is tested.

Let us now assume an experiment aiming at studying the combined effects of different dietary plant proteins and water temperatures on fish growth, with three different dietary plant proteins (soybean meal, corn gluten meal and black eye pea meal) and two water temperature levels (20°C and 30°C). The former represents a qualitative factor showing no structure, while the latter represents a quantitative factor. If the study is conducted following a completely randomized factorial design, there are six possible treatments (Table 4.1). In this experiment, the number of planned contrasts equals the number of degrees of freedom, which is 6 treatments −1 = 5. The ANOVA will verify whether or not the following sources of variations significantly affect fish growth:

1. Interaction protein source × water temperature,
2. Main effect of water temperature,
3. Main effect of protein source.

To interpret ANOVA results, the first step is to check whether or not the interaction protein source × water temperature is significant. If this interaction is significant, only the interactive effect of the two factors is considered and the two main effects are not considered; hence, the interpretation of the data is that every dietary protein source behaves differently at each water temperature. Hence, the effects of water temperature and dietary protein source on fish growth cannot be separated. In contrast, if the interaction is not significant, then the effects of water temperature and dietary protein source are independent and can be analyzed separately. If the effect of water temperature is significant, then the water temperature leading to the highest growth is the best rearing temperature. If the effect of protein source is significant, then the following contrasts can be considered:

1. Soybean meal versus corn gluten meal
2. Soybean meal versus black eye pea meal

Furthermore, if neither the interaction effect nor the main effects are significant, then it is concluded that the individual effect (main effect) and combined effect (interaction effects) of these independent variables do not lead to any change in fish growth.

For details on the formulation, statistical programing and interpretation of orthogonal and polynomial contrasts in planned experiments in agricultural sciences, readers are referred to the book edited by Steel et al. (1997).

4.4.5 Results

The results of aquaculture research are generally expressed in terms of some trait of economic interest such as growth (percentage weight gain, final weight, thermal unit growth coefficient, etc.), feed utilization (feed intake, feed conversion ratio, feed efficiency, protein efficiency ratio, etc.), condition factor (CF) and fish survival. However, reporting other parameters such as fish composition (protein, lipid, amino acids and fatty acids) and disease resistance are common. Anything that went wrong during the study should be mentioned in the Results. The results are presented in the form of text, tables and figures. The text of the results section is generally short and straightforward, reported without discussion. Tables and/or figures help the reader visualize the findings. The text should help the reader interpret data presented in the tables and figures, for example (from Yossa et al. 2014):

> Final body mass was significantly lower in fish fed the biotin-unsupplemented diet than in fish fed diets supplemented with biotin ($P<0.05$) (Table 4.2). The highest growth was observed in fish fed diet containing 0.514 mg biotin kg^{-1}, although not significantly different from fish fed diets with higher biotin levels. Significant differences in growth among treatments were already observed by day 28, with the group fed the biotin-unsupplemented diet showing the lowest body mass (Figure 4.1).

Figures and tables should likewise complement and not repeat either each other or the text. Figures can be photographs, maps, diagrams, graphs or drawings. The format for photos (JPEG, TIFF, etc.) is dictated by the journal. Resolution of photos is normally >300 dpi. Data reported in tables and figures are usually accompanied by a summary of statistical analysis. For most purposes, detailed ANOVA reporting of sums of squares, etc., adds little to the interpretation of results. Most researchers simply include an asterisk or superscript letter or number to indicate which results are statistically significant at a given level of uncertainty (*P*).

All tables and figures must be cited in the text and numbered according to their order of appearance in the manuscript. Table and figure captions should be self-explanatory, meaning that the reader should be able to understand the table and figure without referring to the text. Table captions are generally at the top of the table, while figure captions appear under the figure. Nobody knows why. In a typical aquaculture experiment, for example, a table caption might read: "Weight gain (g) of juvenile (15 g) Nile tilapia (*Oreochromis niloticus*) reared for 90 days in earthen ponds on a prepared diet (T1) or dried bugs (T2). Values with different superscripts within a column are significantly different ($P < 0.05$)".

4.4.6 Discussion

As with the Introduction, the Discussion section of a peer-reviewed aquaculture manuscript is nowhere near as thorough as would be the case for a thesis or dissertation. A Discussion section that goes on for more than three pages usually needs careful review to eliminate things that a professional reader either should know or can easily find in the references cited. The basic point of the Discussion is to put the results into context and explain what they mean to readers and to the development of aquaculture. This is usually done by comparing the findings of a research project with those of others that have done closely related works. Attention should be focused on previously reported findings that do not support the new results. Also of relevance are errors or accidents in the research protocols or execution that make the findings difficult to understand.

Considering that aquaculture is an applied science, the Discussion is not just a list of all the findings of all the articles that have ever been written on a particular topic.

The literature discussed should be of direct relevance to the topic at hand. For example, a paper about essential amino acids in tilapia diets should limit the discussion to other work on amino acids in tilapia and should not get into either other nutrients (e.g., fatty acids) or species with which tilapia have little or nothing in common (e.g., barracuda). It is often helpful if, at the beginning of the Discussion, the objectives and hypothesis are repeated. In this manner, the Discussion gives the authors the opportunity to demonstrate how their work contributes to the advancement of aquaculture and aquaculture science. To keep ideas straight and easy to follow, each paragraph should contain only one idea, and all the paragraphs follow a logical order established back in the Introduction (Shubrook et al. 2010). At the end of the Discussion section, the conclusions and/or recommendations are stated relative to the hypothesis of the study and a closing statement made on the directions for future work. The take-home message should be briefly and clearly stated more or less as it appears in the abstract.

4.4.7 Acknowledgments

This section typically begins with a word of thanks to people and institutions that supported the work, including reviewers, mentors and sometimes family who gave up time so that the work could be successful. For externally funded research, it is important to mention the name (and project number) of the institution that provided the funds used to carry out the experiment.

4.4.8 References

Some scientists seem to think that the more references there are in a paper, the better. While a thorough review of the literature is important in theses, dissertations and review articles, this is generally not the case for original research. For the research to be relevant, most of the articles cited should ideally not be older than 10 years and even less in fast-moving fields such as molecular genetics. The role of references in the text is to provide a context in which the research was carried out (history of the field), save space (standard methods) and give credit where it is due (avoid plagiarism). Citation is not necessary for statements of the obvious or common knowledge. Stating that fish live in water, for example, does not require a reference to some basic biology textbook. A clue as to what should be taken as common knowledge and what needs referencing can be gleaned from the publication date. Anything older than 30 years should either have been accepted as known or rejected as erroneous, so is usually not worth citing. The exact format for citations and references differs among journals. Again, no one knows why. Increasingly, websites are used as citations. Personal communications are a special type of reference for which each journal has a special format.

4.5 Summary of Tips for Good Scientific Writing

- Keep it simple; avoid long sentences that confuse readers; sentences of a maximum of 20–25 words are easy to read and to understand;
- Don't repeat yourself; say things once. If the text is well constructed and easy to follow, it will not be necessary for readers to search around for items of special interest;
- Keep topics in order; this will also make it easy for the reader to follow and find things;

- Use the "grammar and orthography" option of your editing software to correct basic spelling and grammar errors;
- Avoid plagiarism by always appropriately giving credit to the authors of the ideas borrowed from literature or personal communication by adequately citing the sources;
- The research team and trusted colleagues should critically review the manuscript prior to submission;
- Now, go out there, write and write well.

5

Manuscript Submission

Writing the manuscript is one thing, publishing it is another thing.

Once the main draft of the manuscript is completed and all the co-authors are happy with its content and presentation, the authors have to submit it for publication to a receiving journal. Submitting a peer-reviewed aquaculture manuscript is a nonscientific endeavor that calls upon the use of strategic decisions to increase the chances of having the manuscript accepted for publication. This includes decision on choosing the receiving journal (outlet), formatting the manuscript for submission and making sure to avoid immediate rejection of the manuscript.

5.1 Choice of Outlet

Some thought should go into the choice of journal to which the research will be submitted for publication prior to the completion of the manuscript writing. Some aquaculture journals have a regional focus (*North American Journal of Aquaculture, Tropicultura,* etc.); other journals concentrate on certain aspects of aquaculture such as fish health (*Journal of Fish Diseases, Fish and Shellfish Immunology,* etc.), nutrition (*Aquaculture Nutrition*) and economics (*Aquaculture Economics and Management*). Some journals deal with both basic and applied research (*Aquaculture, Aquaculture Research, Journal*

DOI: 10.1201/9780429322648-5

of the World Aquaculture Society, Aquaculture International, etc.) and some with more applied research (*Journal of Applied Aquaculture, Applied Phycology,* etc.).

One should be as specific as possible in aligning the scope of the manuscript to the scope of the targeted journal so as to avoid automatic rejection. More prosaic things to consider in selecting a journal include impact factor of the journal, the duration of the review process, publication fees, reputation of the editorial board and the reputation of the publishing house holding the journal. In a situation of multiple investigators/authors conducting research, all the co-authors should be involved in the choice of the receiving journal prior to submission. In this way, situations like conflicts of interests can be either prevented prior to or declared during the submission process.

5.2 Preparing/Formatting the Manuscript for Submission

Once the manuscript has been constructed and the receiving journal selected, all of the research team should carefully and critically review the manuscript a last time prior to submission. This is to make sure that a best effort has been made to catch mistakes that can make the research team and institute look bad and waste the time of reviewers and Editors. Non-native speakers should generally seek the help of a native English speaker or professional translator to look at the manuscript prior to submission. Many good manuscripts are rejected only because the language is bad. Submitting an article that is full of typographical errors, misspellings, missing references, etc., often earns an immediate rejection. All new computers have spelling and grammar checkers. Use them!

Most of the academic journals use online manuscript submission systems to manage manuscripts from submission to final decision. The submitting author is generally requested to provide a cover letter on behalf of all the co-authors that presents in one to two sentences the main finding of the paper and declares that all the co-authors have agreed on the submission of the manuscript to this journal, that the information contained in the manuscript is original and that the manuscript has not been submitted somewhere else for publication. The cover letter should not exceed half a page. During the manuscript submission, the submitting authors also provide a list of two to five preferred potential peer reviewers as well as a list of non-preferred peer reviewers.

Most journals require that tables, figures and captions appear separately from the main text. Nobody knows why. In this case, each table or figure is presented on a new page. Figure legends usually appear on a separate page, after the references.

5.3 Why Some Submitted Research Manuscripts Are Immediately Rejected

Each journal disseminates scientific information generated by researchers to meet the needs of varied audiences including the industry, policy makers and other scientists working in a specific discipline. Each journal has specific "aims and scope", which present the topics of interest for the journal, as well as the extent of scientific requirements (basic or applied science) for submitted manuscripts. It is essential that each submitted manuscript conforms to the "aims and scope" of the receiving journal, as this represents the initial step in manuscript evaluation. This initial step in manuscript screening is performed by the Editor-in-Chief of the receiving journal. Should a manuscript not conform to the "aims and scope" of the receiving journal, it will immediately be rejected. The "immediate or early rejection" consists of denying a newly submitted manuscript to enter the peer review process. Authors of original manuscripts that are immediately rejected prior to entering the peer review process usually express their unhappiness, disappointment and frustration to the Editors. Some of these authors even inquire about additional evidence that support the Editors' decision not to approve their manuscript to enter the peer review process. This is unnecessary, because the decision to immediately reject the manuscript saves time both to the authors and the editorial board.

A manuscript that complies with the "aims and scope" of the receiving journal successfully passes through the initial screening and then is checked for conformity with "the instructions for authors" of the receiving journal. Again, this second manuscript screening stage is performed either by the Editor-in-Chief or a Managing Editor or an Assistant Editor at the receiving journal. Here, there are four criteria dictating the decision to "approve, reject or unsubmit" the manuscript, including whether or not the manuscript is complete, all identifying information of the co-authors has been removed from the manuscript text, all the co-authors are listed in the online system used to manage the manuscript and the citation and reference styles are correct.

5.3.1 Is the Manuscript Complete?

All the parts of the manuscript including the abstract, materials and methods, results, discussion, references, tables and figures should be prepared strictly in accordance with the "instructions for authors" of the receiving journal and then uploaded through the specified online manuscript submission and management system.

5.3.2 Has All Identifying Information of the Co-authors Been Removed from the Manuscript Text?

For journals that follow a double-blind review system (i.e., the authors and the selected reviewers of a specific manuscript do not know each other's identities), all authors' identifying information must be removed from the manuscript's main text, tables, figures, and illustrations. Only the title page of the manuscript, which is not shared with the reviewers, contains the authors' names, addresses and e-mail addresses, as well as the acknowledgments.

5.3.3 Are All the Co-authors Listed in the Online Manuscript Submission System?

This criterion only applies if there is more than one author for the manuscript. In order to confirm that all the co-authors, in addition to the submitting author, are aware of and agree with the submission of their research results for publication in the receiving journal, the submitting author must provide the full names, addresses and e-mail addresses of all the co-authors during manuscript submission.

5.3.4 Are the Citation and Reference Styles Correct?

All the references, both in the text and in the references list, must be cited strictly in accordance with "the instructions for the authors" of the receiving journal. Also, there must be a perfect match between the references cited in the text and those edited in the reference list. In other words, all the references cited in the text must be edited in the reference list, whereas the reference list must only contain the references that have been cited in the text.

During this second manuscript screening, if the Editors of the receiving journal answer "NO" to any one or more of the previous four questions (criteria), then the manuscript is automatically "unsubmitted". "Unsubmitting a manuscript" means the manuscript is sent back to the authors for appropriate corrections prior to resubmission. Therefore, an unsubmitted manuscript is not rejected but needs to be thoroughly corrected to meet the requirements of the receiving journal. Upon completion of the changes requested by the Editors, the authors resubmit the manuscript, which is checked once again for conformity by the Editors of the receiving journal.

If the Editors answers "YES" to all the previous four questions (criteria) during the second manuscript screening, then the manuscript is automatically "approved". The "approval of a manuscript" allows it to enter the peer review process and to be considered under review. The peer review process involves the scoring of the manuscript by an Associate Editor and at least two peer reviewers and final decision (accept, reject, requires revision or requires editing) by an Editor-in-Chief. When a manuscript undergoes the peer review process and the Editor-in-Chief's decision is either "manuscript requires revision" or "manuscript requires editing", the authors have the option to prepare a revised version of the manuscript. This revised manuscript is thereafter uploaded through the manuscript submission system and is once again subjected to a screening by the Editors, on the basis of the four criteria presented previously. However, there is an additional screening criterion which is applied on revised manuscripts only, which is whether or not the changes made on the initial manuscript have been highlighted in the revised manuscript.

5.3.5 Are All the Changes Made on the Initial Manuscript Highlighted in the Revised Manuscript?

This fifth criterion allows the reviewers to easily see the changes that have been made on the initial version of the manuscript by the authors to produce the current revised version of the manuscript. The highlighting of the changes made in the initial manuscript that appear in the revised manuscript are done using either the "track-changes" mode in Microsoft Word or by using bold or colored text. During this additional conformity check on a revised manuscript, if the Editors answer "NO" to the previous question, then the revised manuscript is automatically unsubmitted and sent back to the authors for additional corrections. If the Editors answer "YES", then the revised manuscript is approved and reenters the peer review process.

6

Reviewing a Manuscript for a Peer-Reviewed Journal

Each author of peer-reviewed articles should reciprocally accept the invitations to benevolently serve as reviewer on other authors' manuscripts.

The main content of this section was first published by the author as a technical article in *World Aquaculture* magazine in 2015 (Yossa 2015). Publishers of many academic journals have produced electronic documents and videos that explain general concepts related to the peer review process and present general expectations of reviewers by the publisher. Although this general information is readily available to the public, especially to authors and reviewers, specific peer review requirements of a particular scientific discipline, such as aquaculture, are scarce in the scientific literature (Babor et al. 2008). For instance, apart from the author's previous paper (Yossa 2015), there is currently no published document that informs aquaculture scientists and researchers on how to perform the review of an aquaculture manuscript. Hence, less-experienced or younger aquaculture scientists, albeit with specialized and highly demanded scientific expertise, might decline an invitation to review a manuscript because they do not know exactly how to do it. Moreover, upon accepting an invitation to review an aquaculture manuscript, the lack of training about peer review and the lack of available specific reviewer guidelines in aquaculture science may lead to the inadequate "one-sentence", useless review occasionally

DOI: 10.1201/9780429322648-6

made by some peer reviewers. The objective of this chapter is to equip aquaculture scientists, researchers and students with guidelines to effectively evaluate manuscripts when they are invited to serve as a peer reviewer for academic journals. Here, I start with a description of the peer review process and then present the principles of manuscript review in aquaculture research. Finally, I provide the qualities sought in a peer reviewer.

6.1 Editorial Board and Peer Review Process

The peer review process (Figure 6.1) involves direct and indirect interactions between the authors of a manuscript, the journal's editorial board (EB), the peer reviewers and the publisher of the journal. The publisher owns the journal and publishes accepted manuscripts. The Editor-in-Chief, who manages the journal and the EB and makes final decisions on manuscripts, is selected on the basis of his/her expertise, experience and reputation in aquaculture research. Associate Editors are selected by the Editor-in-Chief on the basis of expertise in specific subjects that constitute the scope of the journal such as aquatic physiology, larviculture and live foods, aquaculture nutrition, fish health and water quality. The Editor-in-Chief and the Associate Editors represent the EB of the journal. The Editor-in-Chief selects as many Associate Editors as needed so that members of the EB cover all subjects within the journal's scope. In general, the Editor-in-Chief and one Associate Editor handle each manuscript within the journal. However, in certain cases, such as invited manuscripts, the Editor-in-Chief may handle a manuscript alone; as such, the Editor-in-Chief also serves as Associate Editor. Associate Editors, who manage and make a recommendation on individual manuscripts,

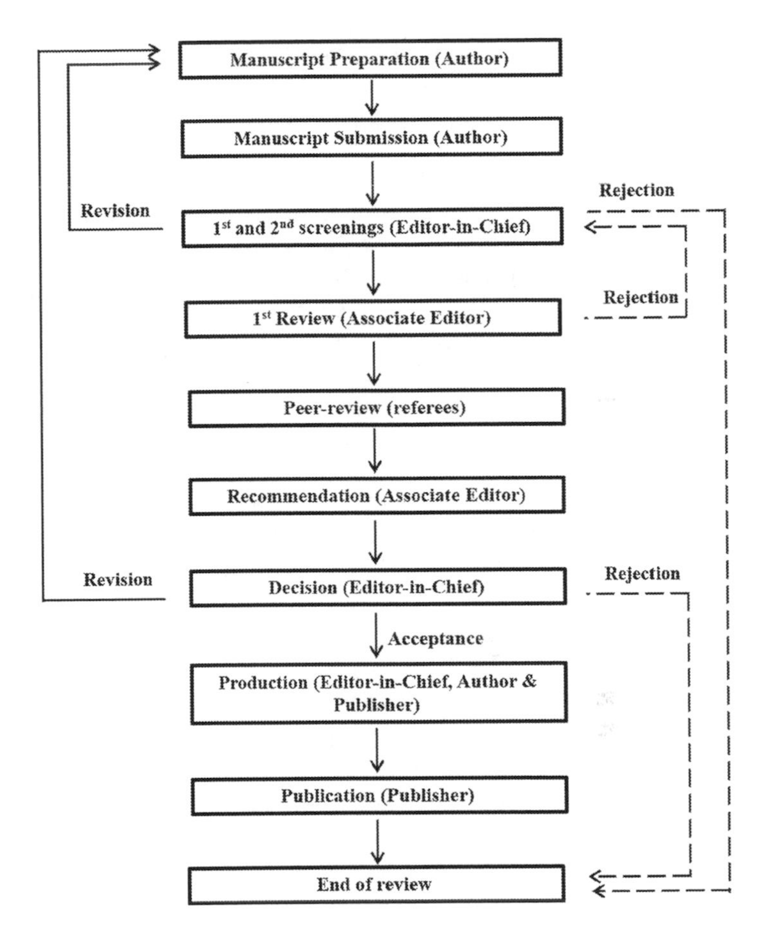

FIGURE 6.1
Schematic representation of the peer review process using an electronic manuscript management system.

select and invite peer reviewers on behalf of the journal. Peer reviewers, also called referees, are anonymous, eminent scientists, who are familiar with the subject presented in a submitted manuscript and who can provide thoughtful critiques on the manuscript.

The peer review process starts with the proper preparation and submission of the manuscript (Figure 6.1). The manuscript must be written and presented strictly in accordance with instructions for authors of the journal and should match the journal's scope (Yossa 2014). Although some academic journals continue to accept hard copies of manuscripts, most aquaculture journals use electronic, online systems to manage the peer review process. In this case, the authors submit an electronic version of their manuscript package, including the main text, tables, figures, figure legends and cover letter through the journal's peer-review management system online. When a manuscript is submitted, the Editor-in-Chief receives an alert that a new manuscript has entered the journal and an initial and second screenings (see Chapter 5) are performed (Figure 6.1). These screenings are articulated around the following criteria:

- Matching of the manuscript subject with the journal scope. The Editor-in-Chief usually reads the title and abstract of the manuscript.
- Completeness of the application package. Here the Editor-in-Chief checks that the main document contains the title, abstract, introduction, materials and methods, results, discussion, conclusion and list of cited references and that tables and figures are properly formatted in accordance with author instructions for the journal.
- Inclusion of the full names, affiliation and contact information, including e-mail addresses, of all co-authors in the electronic manuscript management system.
- Absence of all personal information (names, affiliation, contact information, etc.) of the authors in the manuscript. This is to keep the identity of

authors anonymous to peer reviewers to avoid potential bias in peer review.

- Correct listing of references in the main text of the manuscript and the reference list, citation style strictly in accordance with author instructions for the journal.

If the first criterion is not satisfied, then the manuscript is immediately rejected and it does not go further in the review process. If one or more of the other four criteria are not satisfied, then the manuscript is sent back to the author for revision and resubmission. If all the criteria are satisfied, then the manuscript enters the peer review process and is sent to an Associate Editor (Figure 6.1).

The Associate Editor performs the first review of the manuscript by skimming through the entire manuscript to evaluate if findings contained in the submitted manuscript merit publication. The manuscript is immediately rejected if the Associate Editor finds serious flaws in the manuscript; for instance, if the study is not original, if it does not add anything to the existing knowledge on the topic, if a similar study has been published by other authors elsewhere or if there is a relevant problem in the experimental design or materials and methods such as the absence of statistics to compare treatments. If the Associate Editor finds the manuscript relevant, then peer reviewers are selected and invited to review the manuscript (Figure 6.1).

The Associate Editor may consider the list of preferred and non-preferred potential peer reviewers suggested by the authors during manuscript submission (Chapter 5) or solicit other scientists to serve as peer reviewers. Peer reviewers carry the greatest burden in the peer review process because they are expected to benevolently perform a thorough and timely evaluation of the submitted manuscript. The peer reviewers' task is described in detail in the next section. Upon completion of the review, each peer reviewer submits to the Associate Editor a report that contains comments and a recommendation to either accept, reject, major revision or minor revision (minor editing; Figure 6.1). However, although these are the general choices of recommendations offered by most academic journals, the exact terminologies used in decision-making for each journal are dictated by the publisher and can vary among journals; for instance, recommendations may also include accept after minor revision, accept after major revision and tentatively accepted. Recommendations from at least two peer reviewers are generally required by the Associate Editor to make his/her own recommendation to the Editor-in-Chief (Figure 6.1).

Usually, when both peer reviewers make the same recommendation, the Associate Editor tends to make a similar recommendation to the Editor-in-Chief. The Editor-in-Chief usually relies on the Associate Editor's recommendation to make the final decision on the manuscript. If the Associate Editor's recommendation is to accept, the Editor-in-Chief carefully reads the manuscript before making a final decision, which is usually either minor additional revision (to polish the manuscript) or acceptance. In any other case, the Editor-in-Chief usually conforms to the Associate Editor's recommendation.

When the Editor-in-Chef accepts a manuscript, the publisher generates a manuscript proof. This document is sequentially edited and proofread by the publisher and the authors. With most journals, upon completion of proofreading, an early version of the manuscript is published online as an early view version. This version of the accepted manuscript has not yet been assigned a volume, issue or page range of the journal but bears a digital object identifier (DOI). The DOI is a unique code that identifies a specific electronic document in the universe of scientific papers. An early view paper with a DOI can be cited just in the same way as a published paper. Later, the final version of the early view paper is published in a specific volume, issue and page range, with the same DOI.

6.2 The Principles of Manuscript Review in Aquaculture Research

As an applied science, aquaculture research uses principles and theories of aquatic biology and agriculture to study the production of aquatic animals and plants. However, because it is fairly difficult to draw a line that separates aquaculture science from aquatic biology, some journals that accept aquaculture manuscripts, here referred to as aquaculture journals, are more "discipline-oriented" (genomics, bacteriology and virology, epidemiology, genetics, physiology, conservation science, etc.), while others are more "practice-oriented" (aquaculture, aquaculture management, production sciences, aquatic nutrition, etc.). The goal is publication of new knowledge that strengthens and broadens practical options to maintain and manage aquatic organisms in culture environments. Hence, typical manuscripts submitted to aquaculture journals should contain information that contributes to the growth, survival and welfare of farmed aquatic organisms; the profitability of aquaculture production and/or the sustainability of aquaculture practice.

A letter of invitation to review an aquaculture manuscript sent from the Editor-in-Chief or the Associate Editor of an aquaculture journal to a potential peer reviewer usually includes the title and abstract of the manuscript (and sometimes the name of the submitting author). When the journal uses an electronic manuscript management system, this letter of invitation is accompanied by three web links that allow the invited potential peer reviewer to register a response to the invitation to perform the review by choosing between agree invitation, decline invitation and unavailable. Thus, the first step in a peer reviewer's contribution is evaluation of the manuscript's title and abstract. Reading these allows the peer reviewer to check alignment of the subject of the paper with her/his field of expertise and experience. In the affirmative, the invited scientist might agree to review the manuscript; in the negative, the invitation to review is declined and the Editor-in-Chief or Associate Editor will have to invite another potential peer reviewer. When the invited scientist does not have time to dedicate to this task, he/she selects unavailable and is usually requested to suggest potential alternative peer reviewers for this task. Moreover, the clarity of the manuscript title and editing and presentation of the abstract play a crucial role in helping an interested peer reviewer decide whether or not to agree to review a manuscript. Authors are encouraged to pay special attention to the title and abstract of their manuscript because these two sections of the manuscript are the first parts that are read by the Editor-in-Chief and potential peer reviewers (Yossa 2014).

If the invited scientist agrees to serve as peer reviewer and to proceed with the review of the manuscript, then she/he is requested to create an account on the website of the manuscript management system, where access is given to a pdf file of the complete manuscript package. This account will be used by the peer reviewer to submit the detailed review report accompanied with a recommendation. To preserve anonymity, the peer reviewer usually does not make comments directly on manuscript files but rather in the review report. However, although it is not yet common practice, some journals allow the reviewer to add comments directly to the manuscript file, while making sure to preserve anonymity of the peer reviewer. The peer reviewer may discuss this possibility with the Editors.

The review report comprises two distinct parts: confidential comments to the Editor and major and minor comments to the authors. Each part should start with a brief summary of the peer reviewer's understanding of the research problem and the methodology used to achieve the research objective presented in the manuscript. Confidential comments to the Editor allow the peer reviewer to provide Editors with important information that they do not want to disclose to the authors; for instance, if the peer reviewer has realized while evaluating the manuscript that the content is out of her/his own area of expertise. Major comments to the author consist of general impressions and strengths and weakness of the manuscript, especially regarding quality of the science. Minor comments to the author constitute a point-by-point analysis of the manuscript, from the title to the reference list. To prepare comments to the authors, the peer reviewer thoroughly reads and analyzes the manuscript, starting with the introduction through the conclusion and ending with the title and abstract.

6.2.1 Analyzing the Introduction Section

The peer reviewer evaluates if the subject is original, the research problem is relevant to the aquaculture community and the hypothesis and objectives are well formulated.

6.2.2 Analyzing the Materials and Methods Section

The peer reviewer examines the experimental design, treatments, parameters of interest, sampling, analytical and statistical methods (Yossa & Verdegem 2015), duration of the experiment and materials used to determine the relevance of these to achieve the research objectives.

6.2.3 Analyzing the Results Section

The peer reviewer checks if all response variables and parameters of interest for the study are well presented and properly interpreted in the text, and the tables and figures are clear and standalone so that no additional text is required to understand them.

6.2.4 Analyzing the Discussion Section

The peer reviewer assesses if (1) the results of the study are discussed to confirm or refute the research hypothesis; (2) the results are compared to those of similar or related studies; (3) explanations are based on scientific evidences and concepts are made to support relevant results and discrepancies; (4) the potential weaknesses of the study are presented and discussed; (5) economic, environmental and/or social benefits of the findings in relation to aquaculture are discussed and (6) the pertinent questions for future research are presented for the advancement of science.

6.2.5 Analyzing the Conclusion Section (or End of the Discussion Section if No Conclusion Section)

The peer reviewer checks if authors relate findings to the research hypothesis and present the main conclusion (take-home message) and directions for future research.

6.2.6 Analyzing the Title and Abstract Sections

Upon completion of the reading of the main text, the peer reviewer examines if the manuscript title adequately reflects the study and the findings. The abstract is scrutinized to check if it contains study objectives, the aquatic species of interest, treatments and experimental design, duration of the study, main results and the take-home message. The abstract length should align with the word limit set by the journal.

Overall, as the review progresses, the peer reviewer evaluates the presentation of ideas in a logical and structured order from abstract to conclusion. The peer reviewer also examines if there is a perfect match between references cited in the text and those in the reference list. Although it is not the peer reviewer's responsibility to check spelling and grammar in a manuscript, they can still make suggestions to authors to upgrade the quality of the language expression. Very poor quality of the language in a manuscript can be a sufficient reason to recommend rejection (Yossa 2014).

On the basis of the main criteria earlier and other criteria specific to the aquaculture research subdiscipline, the peer reviewer prepares a report and the decision and score on the manuscript. This information is submitted to the journal, usually within a few weeks following the agreement to review a manuscript. However, the peer reviewer can request an extension of the review period to the Editors. While submitting her/his report, the peer reviewer also indicates her/his willingness to review a revised version of the manuscript if the Editor-in-Chief decides that further revisions are necessary prior to accepting the manuscript. If a peer reviewer recommends further revision, it is appropriate to indicate willingness to review the revised manuscript to assure that the peer reviewer's comments were thoroughly considered in the revision.

6.3 Qualities Sought in a Peer Reviewer

Peer reviewers benevolently make an essential contribution to science in general and the inviting academic aquaculture journals in particular. In an ideal world, each scientist should serve reciprocally as author of her/his manuscript and peer reviewer of other authors' manuscripts. To effectively perform peer review, a reviewer should satisfy some, if not all, of the following standards:

- Be familiar with the subject of the manuscript.
- Be available. If a peer reviewer does not have 3–6 hours to dedicate to review an article within

3–4 weeks of agreeing to the review, the invitation to review should be declined to avoid the inconvenience of delaying the review process.

- Work with integrity. In case of real or potential conflicts of interest, the peer reviewer should inform the Editors. The peer reviewer should be respectful to the author and maintain confidentiality.
- Work with objectivity. The peer reviewer should make comments on the basis of scientific evidence and principles and not on the basis of personal feeling, apprehension and opinion.
- Be accountable for the review. If the peer reviewer is helped in the review by a student, postdoc or colleague, the invited reviewer should read and approve comments prior to submitting the review report on the manuscript.
- Be willing to review a revised version of the manuscript, irrespective of the recommendation submitted.

7

Preparing Authors' Responses to Reviewers' and Editors' Comments

The authors should bear in mind that in the peer-review process the "power" is in the hands of the reviewers and Editors.

When the authors receive the decision from the Editor-in-Chief requesting the revision of their manuscript prior to its resubmission, it is a positive sign that the manuscript has caught the attention of both the Editors (Editor-in-Chief and Associate Editor) and peer reviewers. However, the fact that the manuscript was not rejected does not mean that it cannot be rejected in the next peer review round. It just means that the authors have another chance to improve their manuscript through either better analysis, the presentation and interpretation of the results, the addition of more information or the deletion of irrelevant information. The revised manuscript has to be resubmitted to the receiving journal. In the process of submitting the revised manuscript, there is a space where the authors can either paste or attach the authors' responses to reviewers' comments. Thus, during the manuscript revision phase, the authors have to perform two main tasks, including the revision of their manuscript and the preparation of the authors' responses to the reviewers' comments.

DOI: 10.1201/9780429322648-7

7.1 Revising the Manuscript as per the Reviewers' and Editor's Requests

The authors have to be grateful for the time the reviewers and the Editors spend reviewing their manuscript and express their gratitude through a thorough consideration of all the comments received. In revising the manuscript, the authors have to keep in mind that the reviewers and Editors have the last word in judging whether or not the authors have thoroughly addressed their concerns and whether they would accept the revisions or not. Therefore, the authors have to perform a point-by-point revision of the manuscript and highlight any changes both in the manuscript package (main text, tables, figures, etc.) and in the response to the reviewers and Editors. The authors have to highlight the changes in the manuscript package following the instruction for authors of the receiving journal, which could recommend either the use of track changes or color highlighting to capture the changes made to the manuscript package.

Revising the manuscript package does not mean that the authors have to agree with all the comments made by the reviewers. It can happen that reviewers' (and even Editors') comments are not relevant: nobody knows it all. In this case, the authors do not make the change in the manuscript package but prepare a strong rebuttal to explain and justify their opposition to the comment in the author's response to the reviewer's and Editors' comments.

7.2 Responding to Reviewer's and Editors' Comments

The authors should prepare a response to each and every comment made by the reviewers and the Editors. These responses have to be aligned with the point-by-point changes made and highlighted in the manuscript package. The responses from the authors have to be written using a polite and respectful style that includes words such as "please" and "thank you" even if the authors consider that the reviewers' comments were not done in a polite and respectful manner. At this stage in the peer review process, the power is in the hands of the reviewers and Editors, and although what I call "power" here can be considered unethical, it is part of the peer review process, just like in any human endeavor where some people have to "judge" other people's work. In practice, the authors should first number the reviewer's comments, then number their response accordingly. Here are some fictive examples:

Comment 1: The authors should revise the abstract and properly define the treatments.

Authors' response 1: Thank you for your comment. We have revised the abstract carefully and defined the treatments as per your comment. Please find the change highlighted in yellow in the main document.

Comment 2: The authors should reanalyze the data using a multiple comparison test.

Authors' response 2: Thank you for your comment. We have carefully considered your comment, but we would like to keep our statistical analyses as they are for the following reasons: (1) the independent experimental variable in this study is quantitative (water temperature) with six levels (20°C, 22°C, 24°C, 28°C, 30°C, and 33°C); therefore, the only meaningful statistical test that can help us estimate the optimal water temperature to grow the fish is a regression; (2) the relevance of using regression to analyze the data of experiments like this has been discussed in Yossa and Verdegem (2015) and used in many other recent publications including Dawood et al. (2020): https://doi.org/10.1016/j.aquaculture.2019.734571; and Kottmann et al. (2020): https://doi.org/10.1016/j.aquaculture.2020.735581. Therefore, we would appreciate if the reviewer would kindly let us keep the data analyses in the manuscript as they are.

Note that the first example (comment 1 and Authors' response 1) is a case where the authors accept the reviewer's comment and make the change accordingly, while the second example (comment 2 and Authors' response 2) is a rebuttal to reviewer's comment. However, it is important to avoid submitting a rebuttal to every comments made by the reviewers, although it is highly tempting to do so. Preparing a rebuttal for each comment is like starting a fight with many people, who are individually much stronger than you. This also suggests that the authors are not humble enough to consider other people's opinion on their work, which is the essence of the peer review process. Remember that the reviewers have been invited to review your manuscript because they are subject matter experts and their objective opinion on your manuscript should count. One way to avoid opposing each of the reviewer's comments is to read the comments three or four times before starting to address them. In this manner, you have the time to digest the comments, calm down and chose the best angle to either accept or oppose them, using an appropriate formatting.

7.3 Formatting Authors' Responses to Reviewers' and Editors' Comments

In responding to the reviewers' and Editors' comments, the authors should avoid using formatting styles such as italics, underlining, colors and highlighting which often disappear in the manuscript management software, as the manuscript management software does generally not support all these formatting styles. Numbering with common modern numerals (1, 2, 3, etc.), as in the previous two examples is enough and efficient.

8

Writing a Research Report

The writing of a good research report facilitates the writing of the subsequent good peer-reviewed manuscript.

A research report is a document that provides information on the progress of a research project or results of the completed research project and is generally considered as a deliverable or output of a research project. It is either submitted internally to the hierarchy or to the project sponsor on a predetermined frequency, which could be monthly, quarterly, annually or finally, depending on duration of the project and the project management applied by the organization or department receiving the report, also called the sponsor. There are typically two types of research reports: the research progress report and the final report.

8.1 Research Progress Report

The research progress report is produced during the course of the research activity or project, to present the progress made either since the beginning of the research project or the progress accrued since the production of the previous progress report. It shows the progress against the planned deliverable, while presenting the challenges, constraints and lessons that have been observed and recorded as the research is implemented. It also suggests the corrective

DOI: 10.1201/9780429322648-8

actions and the adaptations that the research team are putting in place in order to assure the effective completion of the research project.

8.2 Final Research Report

A final research report is an elaborated document produced upon completion of the research project, in order to present the results of the overall research project and the lessons learned. The research report presents not only the findings generated by the research but also the additional results related to the completion of the research project such as the training of staff (human capacity building), the building of new research facilities (institutional capacity building), the creation and/or strengthening of the partnership among the project team members and between the leading organizations and the sponsor and the dissemination of research findings through participation in conferences, workshop and seminars. Unless otherwise specified by the agency receiving the report, the presentation of the project findings is done following the same structure as the peer review article (introduction, materials and methods, results, discussion and conclusion) and follows the same level of scientific rigor, but with a less specialized jargon, considering that the readers of the research reports are not necessarily scientists. However, the research report usually contains an executive summary, which is inserted at the beginning of the report, just before the abstract. This executive summary presents the purpose of the report, the overall methods used including the source of funding and the teams involved in the research project and the main findings and their relevance. Moreover, in the text of the research report, information is provided on whether the project was completed within the agreed budget and timeline, why not and what actions were taken to align or realign the research project. The lessons learned during the achievement of the research project are listed in the research report. The sponsor may use this information as "field" experience and new risks to be considered while assessing new applications for research funding and for evaluating other completed research projects in the future. The research team uses the lessons learned to make adjustments during the design and implementation of future research projects and to promote best research practices in a given context.

8.3 Data in Research Report versus Peer-Reviewed Manuscript

Ideally, once the final research report is produced and submitted to the sponsor, the next step would be that the authors write a peer-reviewed article based on the results presented in the research report and submit it to an academic journal for publication. However, in reality, the research report often belongs to the sponsor, especially in contract research where the authors act a service provider to a sponsor who commissions and pays for the research project. In this case, the sponsor has the right to either keep the information contained in the research report for her/himself, publish the report as it is on her/his website or any other outlet (maintaining the authorship as the same as in the report) or grant the right to the authors to prepare and submit a peer-reviewed manuscript using the data presented in the research report. It is thus important that the research team and the sponsor agree on who owns the data and the way the data will be disseminated, prior to signing the sponsorship agreement, in order to avoid any conflict at the end of the experiment. In a situation where the agreement is that the sponsor will publish the research report while the authors simultaneously have the right to publish the data in a peer-reviewed manuscript, then it is important that the authors carefully choose the data and the format of the data that they include in the project report. This is because if the data contained in the final research report are published by the organization receiving the report, chances are that these data will no longer be accepted for publication in a peer-reviewed journal. Hence, it is a common practice to present the main trends and the high-level findings in the research report and present the data in detail in the peer-reviewed manuscript. To be safe, upon completion of the writing of the project report, whatever the initial agreement regarding the dissemination of results, the authors should inform the organization receiving the report that they would like to publish part of the information contained in the scientific report in a peer review journal; in this way, the sponsor could kindly either wait until the peer-reviewed manuscript is published before publishing the research report or restrict the sharing of the research report to internal use for a while.

Authors should dedicate time and effort in producing a quality research report, as it is easy to write a quality manuscript from a quality research report. In practice, some parts of the research reports, such as introduction, materials and methods and the results, can easily be copied and pasted in the peer-reviewed manuscript.

9

Writing a Working Paper

A working paper can be a good alternative for information that would not be accepted as it is in peer-reviewed journals.

A working paper is a preliminary version of an academic manuscript that is published while the work on the manuscript is still in progress. A working paper is thus published prior to the completion, submission, review or publication of the academic manuscript. The working paper is published either to share results of the research project (in the case where they are weak and might not be accepted for publication in an academic journal) or to seek feedback from the scientific community prior to completing the work on the academic manuscript. The writing style of a working paper is between the peer review manuscript and the scientific research report. The authors should present and interpret the results as in peer-reviewed papers but are sometimes allowed to include their personal opinion and theoretical assumptions in the discussion section of the working paper.

The working paper is either published as an internal publication by the authors' institution or any other organization accepting working papers in the specific field of the authors. The receiving organization provides a format guideline for the working paper, which is usually simple with sections such as the executive summary, introduction, materials and methods, results, discussion and concluding remarks. Unlike manuscripts that usually do not have a page or word count limit, the working paper is usually limited to 10–15 pages. Working papers are usually not

DOI: 10.1201/9780429322648-9

peer-reviewed, and are cited as gray literature, as it is the case with research reports and technical publications.

Although it is a common practice to publish working papers prior to the academic manuscript in social sciences, this practice is not common in pure and applied sciences such as aquaculture, where academic journals usually only accept manuscripts that contain the data that have not been published anywhere else, except sometimes in conference abstracts. Therefore, authors of aquaculture manuscripts should be cautious when invited to submit a working paper, as the publication of their results in a working paper might prevent from publishing them in a peer review journal later.

10

Writing a Conference Abstract

Writing precedes the presentation at a conference.

Attending local, national and international conferences is an integral part of the professional life of an aquaculture scientist. Aquaculture conferences allow the scientist to meet with her/his peers; get familiar with progress of research in aquaculture; assess recent information on the use of science to advance aquaculture production; learn from other participants' findings and experiences; share one's recent findings and ideas; and meet and socialize with students (the future), colleagues, friends and possible future employers or funders. Although it is possible to attend any conference only by paying conference fees, it is expected that scientists actually fully participate in a scientific conference by presenting their recent findings, either in the form of an oral presentation or a poster presentation. In this case, scientists are invited to submit a conference abstract for review and acceptance, prior to receiving the invitation letter to attend the conference and make a presentation or a poster.

10.1 The Content of the Conference Abstract

Whether the scientist applying for participation in an aquaculture conference wants to make an oral or poster presentation, the requirements for the abstract are the same.

DOI: 10.1201/9780429322648-10

The length and format of conference abstracts vary from conference to conference. Some conferences will limit the number of words to 150, while other conferences will allow the author to use a full page, with specific margins, line spacing and font. In any case, the following information is required in each abstract:

- The title of the abstract,
- Authors' names, with the first name being that of the presenting author,
- The presenting author's address and e-mail, and
- The main text.

Some conferences allow the insertion of figures and/or tables in the main text, while maintaining the allocated space or word count limits. While writing the main text, authors should define abbreviations the first time they appear in the text and make sure all the parts (problem statement, general methodology, main results and take away message) of the abstract are clearly presented (see abstract of the peer-reviewed manuscript in Chapter 4). Sometimes, the deadline for abstract submissions occurs while only preliminary results are available. Nevertheless, the authors must only present the results that are available and well analyzed in the abstract, even if they don't show the complete picture. However, if by the time the conference takes place, the full spectrum of the results is available, the authors can still add these results in their presentation (oral or poster) and say a few words about the relevance of this additional information during their presentation.

10.2 Submission and Review of the Conference Abstract

Once the writing of the conference abstract is complete, it is then submitted to the conference organizer for review by the conference steering and/or scientific committee. It is important that the abstract is submitted using the channel (usually online) suggested by the conference organizer, within the deadline for submitting abstracts, otherwise this abstract will not be considered for review and acceptance. The conference steering and/or scientific committee selects the best abstracts to be presented at the conference, based on the quality of abstract presentation and the relevance of the information contained in the abstract for the advancement of aquaculture science and production. There is usually no limit to the number of abstracts that have to be accepted in international conferences, though in some national and regional conferences only a limited number of participants are allowed and the selection of abstracts is more competitive. While submitting the abstract, the submitting author is usually requested to suggest whether the abstract will be presented in the form of an oral or poster presentation. The presenting author will choose her/his preferred presentation technique. However, when the conference steering and/or scientific committee accepts the abstract, they also decide whether the abstract will be presented as suggested by the submitting author (oral presentation, for instance) or in the other form (poster, for example). This decision is based on the limited number of spots for oral presentations in specific conference sessions, while there is usually no limit in the space for poster presentations.

During abstract submission, the submitting authors are requested to mention whether she/he is a student or not. This is because there are sometimes travel awards for a limited number of abstracts deemed best by the conference steering and/or scientific committee. These best abstract awards are usually given to the student solemnly and sometimes include a parchment, plate and prize money. I received two of them when I was a PhD student at Université Laval, Canada, namely the Younger Scientist Award at the 14th International Symposium on Fish Nutrition and Feeding in China in 2010 and the Best Student Award at the World Aquaculture Conference in Brazil in 2011. These two awards gave a big boost to my scientific career and my personal confidence and also made me famous and important in the eyes of colleagues and potential employers. For instance, at the World Aquaculture Conference in Brazil in 2011, I met the professor who offered me a 2-year postdoctoral training at his laboratory at West Virginia State University in the United States between 2013 and 2015. It is also at this conference that I remet (after 3 years of not seeing) my mentor who wrote the foreword of this book, who was so happy with my award that he offered me the opportunity to work as his Assistant Editor at the *Journal of Applied Aquaculture* for 4 years (2012–2015) before replacing him as the Editor-In-Chief in 2016. It is thus important that students (and their mentors) take the time to produce quality one-page conference abstracts, as such a one-pager can change the life and career of the students for the best.

10.3 Preparing for the Attendance of an Aquaculture Conference

Once the abstract is accepted, the presenting authors are usually invited to pay the conference fees and then are provided with the invitation letter and all the necessary information for visa application and booking of accommodation. At this stage, the main author should start preparing her/his talk, and the supporting PowerPoint (ppt.) or poster presentation. The presentation (oral or poster) should be shared with all the co-authors prior to its finalization. The ppt. presentation is usually turned over to the conference manager the first day of the conference or the day before the session. On the other hand, a poster is printed prior to traveling to the city or country where the conference will be held. It is important that the presenter strictly follows the format of either the ppt. or poster, as any change in the format could lead to an incompatibility for the IT system for presentation or the space reserved for the poster, and the presentation will not be possible. It is also important that the presentation is made in the language of the conference, which could be English or any other combination of language with simultaneous interpretation.

Prior to the start of the conference, a book of abstracts (or abstract book) is published by the organizing committee and distributed to all the participants during the physical registration at the conference venue as part of a conference package. This book contains all the abstracts that were accepted by the conference steering and/or scientific committee, as well as the information on the oral or poster sessions and date and time at which they will be presented. The conference package usually also includes a conference bag, advertisements from sponsoring companies or organizations, pen/pencil and a notebook, in addition to the book of abstracts.

It is important that the presenter is ready to present before traveling to the conference venue, because for any reason, the initial program of the conference can change, and the presenter is invited to present earlier than planned. Also, a well-prepared presenter is more relaxed during the conference and spends most of his/her time learning and networking and needs only minimal time to rehearse a few times prior to presentation at the conference. Personally, I prefer oral presentations to poster presentations, because during oral presentations, people who come to the room only enter when they are interested in listening to my presentation and they are supposed to be sober. My experience with poster sessions is that they often happen during the afternoon and at the same time with happy hours. As such, people are allowed to attend the poster sessions with a glass of beer in their hand, and I find it annoying to explain my poster to people who are half drunk in front of me while I am thirsty.

11

Writing an Article for Conference Proceedings

The information contained in a full-text paper published in conference proceedings can no longer be published in a separate peer-reviewed article, and vice versa.

Once a scientific aquaculture conference is completed, some conference steering and/or scientific committees publish the conference proceedings, either in the form of an internal final conference publication or in partnership agreement with an academic journal that will publish the content of the conference proceedings in the form of individual peer-reviewed articles. In any case, the conference steering and/or scientific committee selects some temporary Editors who will manage the review process of the proceeding articles. Then, in collaboration with these Editors, the conference steering and/or scientific committee invites a selected number of authors of conference abstracts to develop their abstracts into full manuscripts. These full manuscripts are either reviewed by the Editors only and then published in the conference proceeding as regular non-peer-reviewed articles or they follow a strict peer review process led by the Editors and are thereafter published as peer-reviewed articles in an academic journal, in a manner similar to that described in Chapter 4. Although the articles published in each of these cases might be similar in content and format, in the former case the published article is considered as gray literature, while in the latter case, it is considered as an academic paper.

DOI: 10.1201/9780429322648-11

Here again, it is worth noting that any data published in aquaculture conference proceedings as a non-peer-reviewed article can no longer be considered for publication later in an aquaculture academic journal. Therefore, if the authors are planning to publish their results in an academic journal as a peer-reviewed article, they should decline the invitation to publish their findings in a conference proceedings, unless the papers contained in the conference proceedings are published by an academic journal as individually peer-reviewed articles.

12

Writing a Technical Article for a Magazine

A technical article published in aquaculture magazines is a great way to disseminate research results to a larger audience.

It is most difficult for a scientist to write a simple popular article than a complicated peer-reviewed article.

Magazines are published by either professional societies, lobby or advocacy groups or any publisher who intends to put together and disseminate technical information available on aquaculture. There are many aquaculture magazines published in English, including *World Aquaculture, International Aquafeed, Hatchery international, Aquaculture Europe Magazine, Aquaculture Africa Magazine, Aquaculture Magazine, Aquaculture Asia Pacific, AQUA Practical* and many others published in other languages. Magazines usually publish technical/popular articles covering all aspects of aquaculture to support the aquaculture industry. These popular articles, which are considered as gray literature, are usually short, written in simple language to facilitate understanding by the general public and, especially, farmers and other technology users. Even the articles presenting advances in a technical topic use simple language and a story-like approach to convey the message. While writing a manuscript for a magazine, the authors should always bear in mind that the magazine is aimed at reaching the general public and that anyone who can read should be able to understand the message of the

DOI: 10.1201/9780429322648-12

article, without having to open a dictionary. Therefore, academic manuscripts resulting from fundamental studies and written with highly specialized jargon and basic scientific approaches are not appropriate for publication in aquaculture magazines. However, the main message of an academic aquaculture research can be phrased in such a way to cover one or several pages in a magazine.

Many aquaculture scientists find it difficult to write popular articles and only use academic journals and conferences as means of disseminating their findings. This situation is supported by the fact that a gray article is considered less powerful as a scientific achievement in assessing a scientist for promotion, graduation or funding. However, many donor organizations funding applied research to support farmers and policy makers in the aquaculture industry are gradually giving high consideration to popular articles, as an efficient means to disseminate findings from applied aquaculture research. It is likely that in the future, some donor organizations will give as much value to popular aquaculture articles as peer-reviewed aquaculture articles, and the most productive scientists would be those who do not only publish peer-reviewed articles in high impact factor journals but also publish the main messages of their research findings in the form of popular articles in magazines.

The use of pictures and illustrations is highly encouraged in popular articles, "a picture (being worth) worth a thousand words". Interesting pictures are those that show the application of the technique or innovation that is described in the article. It is important to accompany the picture with appropriate self-explanation title and legend.

Here are a few tips for writing a popular article for a magazine:

- Avoid using technical jargon
- Explain each abbreviation when it first appears in the text

- Use simple sentences no longer than 20 words
- Insert pictures and illustrations in the manuscript
- Write a manuscript of 1,000–3,000 words
- Seek the assistance of a copy editor for the improvement of the "simple language"
- Let some friends who are not familiar with the topic covered by the magazine manuscript read and provide their feedback on the way they understand the message.

It is worth noting that magazines also publish all sorts of advertisements. Although all these advertisements could seem annoying for an author, it is good to keep in mind that these advertisements are the main source of income to the publisher of magazines. Moreover, some readers are only interested in the advertisements that show the most recent advancements in, for example, water quality management or in fish feed production, while the publisher is trying to satisfy the expectations of its wide range of readers, who are part of the general public.

13

Overview on Managing a Manuscript Writing Project

Writing a manuscript is a project by itself.

It often happens that one individual aquaculture scientist is invited to lead the writing of a manuscript resulting from collaborative research that she/he conducted jointly with a group of other aquaculture and non-aquaculture scientists or is asked to select a few other scientists to initiate and write a research review article. Each of these cases is an example of the management of a writing project, with the lead author being considered as the project manager. According to the *Project Management Body of Knowledge Guide (PMBOK® GUIDE)* produced by the Project Management Institute, "a project is a temporary endeavor undertaken to create a result, service or product". The words "temporary", "create" and "product" in this definition will be used to present a writing endeavor as a project that has to be managed using the best available project management tools, knowledge and practices.

The word "temporary" shows that a project has a life span, with a start (kickoff date) and an end (closure date). A writing project should thus not last forever. The authors have to keep in mind that science evolves and that if they wait too long, either another research group will publish similar results before them or their findings will no longer be relevant by the time they complete the writing. Depending on the extent of data analysis and the complexity of the topic covered by the aquaculture manuscript, the

DOI: 10.1201/9780429322648-13

number of authors could be so high (sometimes above 50) that it takes years to complete the writing. However, for experiments that have been conducted in 12 weeks, as it is the case in most aquaculture research, it would be reasonable to write the manuscript within 6 months following the completion of the experiment and the laboratory analyses.

The word "create" means the writing team has to produce something from scratch, an aquaculture manuscript, as a collective endeavor. In the process of creating something, actions have to be taken and active participation of the co-authors is required. As the project manager, the lead author of a writing project should lead the team in a coordinated manner and work with them to complete the creation of a quality manuscript in time, using the available budget. In so doing, the lead author has to make sure that the expertise of each co-author is used at the right time and in the right section of the manuscript. For instance, if the paper is dealing with the application of modeling techniques in analyzing nutrient dynamics in aquaculture, the lead writer could make sure that in the co-author team there is expertise in nutrient modeling, aquaculture production systems, environmental sciences, fish nutrition, fish physiology and aquatic ecology. The lead author should make sure that expertise from each of the co-authors is put to use in a coordinated manner to produce a comprehensive manuscript that covers all the aspects of the topic.

The word "product" suggests that at the end of the writing project, there should be a product, a published article, which can be used to close the project. It is important to insist on the fact that the writing project does not stop at the production and submission of the manuscript, as the co-author team can still be invited to provide a substantial writing contribution during the rounds of revisions of the paper, upon receiving the comments from the reviewers.

As in any project, it is important that the lead author maintains a good and regular communication with the co-author team, in order to keep the writing momentum going forward, ensure the quality of the work, streamline the timeliness of the contributions and meet the deadlines agreed upon during the initiation of the writing project. The lead author should also play the role of the corresponding author. As such, she/he is expected to communicate with the other stakeholders involved in the writing project, including the funders of the project, the editors, peer reviewers and publisher of the receiving journal, and the potential end-users of the information that will be contained in the published article.

14

Practical Considerations to Improve Scientific Writing Skills

If you can type the text fast, then you will likely have more time to think.

- Learn how to type with your ten fingers on your keyboard. The ten-finger typing skill will improve your efficiency in writing, as the faster you are in writing, the more time you will have for thinking or writing even more;
- Only consider inviting co-authors who can add something to the writing in a timely manner; "better alone than badly accompanied". Interacting with co-authors can be difficult at times, so don't invite co-authors with whom you cannot normally sit down and have a drink or a conversation, just because you would like to have their names in your published papers or because you want to show that you are a team player;
- Reading improves the writing. As such, please read at least 100 times as much as you would like to write, and you will learn a lot from the writings of others that will help you improve your own writing skills.

DOI: 10.1201/9780429322648-14

15

Conclusion

Now you don't have any excuse not to write either an abstract or a full manuscript after completing an aquaculture research project. The fact that you read this book shows that you want to write, so just go out there and disseminate your findings using the only communication channel that never dies, writing. Remember that the Bible, Koran, etc. have lasted this long because they were written, so if you want your ideas to last forever, then write them down and publish them.

DOI: 10.1201/9780429322648-15

References

Aarssen, L. W., T. Tregenza, T. A. E. Budden, C. J. Lortie, J. Koricheva, and R. Leimu. 2008. Bang for your buck: Rejection rates and impact factors in ecological journals. *The Open Ecology Journal* 1, 14–19.

Babor, T. F., K. Stenius, S. Savva and J. O'Reilly. 2008. *Publishing Addiction Science: A Guide for the Perplexed,* 2nd ed. Essex: Multi-Science Publishing Company.

Bhujel, R. C. 2009. *Statistics for Aquaculture*. Wiley-Blackwell.

Björk, B.-C. (2019). Acceptance rates of scholarly peer-reviewed journals: A literature survey. *El profesional de la información* 28(4), e280407. Doi: 10.3145/epi.2019.jul.07.

Cargill, M., and P. O'Connor. 2013. *Writing Scientific Research Articles: Strategy and Steps,* 2nd ed. Oxford: Wiley-Blackwell.

Carpenter, K. 2001. How to write a scientific article. *The Journal of Paleontological Sciences*. http://www.aaps-journal.org/submission%20pdf/How%20to%20Write%20a%20Scientific%20Paper.pdf.

Chew, V. 1976. Comparing treatment means: A compendium. *HortScience* 11, 348–356.

Davis, M. J. 2010. Contrast coding in multiple regression analysis: Strengths, weaknesses, and utility of popular coding structures. *Journal of Data Science* 8, 61–73.

Dawood, M. A. O. M. Zommara, N. M. Eweedah, and A. I. Helal. The evaluation of growth performance, blood health, oxidative status and immune-related gene expression in Nile tilapia (Oreochromis niloticus) fed dietary nanoselenium spheres produced by lactic acid bacteria. Doi: 10.1016/j.aquaculture.2019.734571.

Day, R. A. 1994. *How to Write and Publish a Scientific Paper*. Phoenix, AZ: Oryx Press.

Elefteriades, J. A. 2002. Twelve tips on writing a good scientific paper. *International Journal of Angiology* 11, 53–55.

Gates, C. E. 1991. A user's guide to misanalyzing planned experiments. *HortScience* 26, 1262–1265.

Gill, J. L. 1973. Current status of multiple comparison of means in designed experiments. *Journal of Dairy Science* 56, 973–977.

Hayer, C.-A., M. Kaemingk, J. J. Breeggemann, D. Dembkowski, D. Deslauriers, and T. Rapp. 2013. Pressures to publish: Catalysts for the loss of scientific writing integrity? *Fisheries* 38, 353–355.

Hengl, T., and M. Gould. 2002. Rules of thumb for writing research articles. http://www.itc.nl/library/papers/hengl_rules.pdf.

Jennings, C. A., T. E. Lauer, and B. Vondracek. 2012. *Scientific Communication for Natural Resource Professionals*. Herndon, VA: American Fisheries Society.

Kottmann, J. S., J. Tomkiewicz, I. A. E. Butts, I. Lund, C. Jacobsen J.G. Støttrup, and L. Holst. Effects of essential fatty acids and feeding regimes on egg and offspring quality of European eel: Comparing reproductive success of farm-raised and wild-caught broodstock. Doi: 10.1016/j.aquaculture.2020.735581.

Little, T. M. 1981. Interpretation and presentation of results. *HortScience* 16, 637–640.

Madden, L. V., J. K. Knoke, and R. Louie. 1982. Considerations for the use of multiple comparison procedures in phytopathological investigations. *Phytopathology* 72, 1015–1017.

Masic I. 2012. Plagiarism in scientific publishing. *Acta Informatica Medica* 20(4), 208–13. Doi: 10.5455/aim.2012.20.208-213. PMID: 23378684; PMCID: PMC3558294.

Montgomery, D. C. 1997. *Design and Analysis of Experiments*, 4th ed. New York: John Wiley & Sons.

NSF (National Science Foundation). 2017. National Center for Science and Engineering Statistics; SRI International; Science-Metrix; Elsevier, Scopus abstract and citation database (https://www.scopus.com/), accessed July 2017.

Olsen, C. H. 2003. Review of the use of statistics in infection and immunity. *Infection and Immunity* 71, 6689–6692.

Petersen, R. G. 1977. The use and misuse of multiple comparison procedures. *Agronomy Journal* 69, 205–208.

Rafter, J. A., M. L. Abell, and J. P. Braselton. 2002. Multiple comparison methods for means. *SIAM Review*. 44(2), 259–278.

Robbins, K. R., A. M. Saxton, and L. L. Southern. 2006. Estimation of nutrient requirements using broken-line regression analysis. *Journal of Animal Science* 84(E. Suppl.), E155–E165.

Robbins, K. R., H. W. Norton, and D. H. Baker. 1979. Estimation of nutrient requirements from growth data. *Journal of Nutrition* 109, 1710–1714.

Russell W. and R. Burch. 1959. *The Principle of Humane Experimental Technique*. London: Methuen & Co Ltd.

Saville, D. J. 1990. Multiple comparison procedures: The practical solution. *The American Statistician* 44, 174–180.

Shubrook, J. H., J. Kase, and M. Norris. 2010. How to write a scientific article. *Osteopathic Family Physician* 2, 148–152.

Steel, R. G., J. H. Torrie, and D. A. Dickey. 1997. *Principles and Procedures of Statistics: A Biometrical Approach*, 3rd ed. Boston, MA: McGraw-Hill Series in Probability and Statistics.

UNESCO. 2010. *UNESCO Science Report: The Current Status of Science around the World*. Paris: UNESCO.

Yossa, R. 2014. Writing a scientific manuscript from original aquaculture research. *Journal of Applied Aquaculture* 26, 293–309.

Yossa, R. 2015. Toward the professionalization of aquaculture: Serving as peer reviewer for an academic aquaculture journal. *World Aquaculture* 46(4), 47–50.

Yossa, R., M. Desjardins, C. S. Pelletier, C. Roussel, C. Landry, and T. J. Benfey. 2018. Erythrocyte length measurement is a valid alternative to flow cytometry for ploidy assessment in Arctic charr *Salvelinus alpinus*. *Aquaculture Research* 49(11), 3601–3605. Doi:10.1111/are.13827.

Yossa, R. and M. Verdegem. 2015. Misuse of multiple comparison tests and underuse of contrast procedures in aquaculture publications. *Aquaculture* 437, 344–350.

Yossa, R., P. K., Sarker, D. M. Mock, and G. W. Vandenberg. 2014. Dietary biotin requirement for growth of juvenile zebrafish Danio rerio (Hamilton-Buchanan). *Aquaculture Research* 45, 1787–1797. Doi: 10.1111/are.12124.

Yossa, R., P. K. Sarker, E. Proulx, and G. W. Vandenberg. 2015. The effects of the dietary biotin on zebrafish Danio rerio reproduction. *Aquaculture Research* 46, 117–130. Doi: 10.1111/are.12166.

Yossa, R., P. K. Sarker, S. Karanth, M. Ekker, and G. W. Vandenberg. 2011. Effects of dietary biotin and avidin on growth, survival, feed conversion, biotin status and gene expression of zebrafish Danio rerio (Hamilton-Buchanan). *Comparative Biochemistry and Physiology Part B* 160, 150–158.

Index

Note: **Bold** page numbers refer to tables and *italic* page numbers refer to figures.

Printed in the United States
by Baker & Taylor Publisher Services